Managing Engineered Assets

Joe E. Amadi-Echendu

Managing Engineered Assets

Principles and Practical Concepts

 Springer

Joe E. Amadi-Echendu
Graduate School of Technology
Management
University of Pretoria
Pretoria, South Africa

ISBN 978-3-030-76053-3   ISBN 978-3-030-76051-9   (eBook)
https://doi.org/10.1007/978-3-030-76051-9

This Springer imprint is published by the registered company Springer Nature Switzerland AG
The registered company address is: Gewerbestrasse 11, 6330 Cham, Switzerland

*To my parents, my siblings, my wife,
my children, and my grandchildren.*

*To Ted Higham, my mentor; yes, I finally
wrote the book long after you have passed.*

Foreword by Georges Abdul-Nour

Complex assets systems are essential to the functioning of our society. They include but are not limited to oil platforms, transportation networks, water and gas distribution systems, nuclear power plants, etc. Utilities must constantly find strategies for remaining competitive in an era characterised by transitions, uncertainties, turbulence, and worldwide competition. Furthermore, they face the challenges of sustainable development and increasingly stringent environmental regulations.

In this increasingly complex world, achieving companies' reliability, resiliency, and robustness criteria requires considering and integrating numerous interacting factors that result from different layers: technical, technological, environmental, social, and organisational.

In order to overcome this complexity, Engineering Asset Management (EAM) has emerged as a vital multidisciplinary area of knowledge and practice consequent upon the vast scope, range, and variety of engineered assets that continue to facilitate all aspects of human existence from earliest times to our current era. Along with the increasing formalisation of management systems comes the need for a curriculum for teaching, research, training, and practice in Engineering Asset Management.

This book, written by one of the most world's leading authors and practitioners in EAM, Dr. Joe Amadi-Echendu, provides a new perspective and a transformative approach to the multidisciplinary scope of Engineering Asset Management. The highest power of the book is that the content includes multidisciplinary principles and practical concepts for teaching, research, training, and asset management. The discourse captivates thinking for academics and practitioners alike.

I highly recommend this book to students, academics, policy makers, and practitioners who wish to embrace innovative ways of managing the vastness of engineered assets in the era of Society 5.0 concomitant with fourth industrial revolution technologies.

Georges Abdul-Nour Ph.D., Ing., FISEAM
Editorial Board of International Journal of Production
Research, Associate Editor, International Journal of Metrology
and Quality Engineering (IJMQE)
http://www.uqtr.ca/inrpme
https://www.mdpi.com/journal/sustainability/
special_issues/risk_asset

Professeur Titulaire
Titulaire Chaire HQ pour la Géstion des Actifs
Academy of Industrial Engineering
Texas Tech University
Lubbock, USA
http://www.uqtr.ca/inrpme
https://www.mdpi.com/journal/sustainability/
special_issues/risk_asset

Foreword by Adolfo Crespo Márquez

Engineering Asset Management has rapidly evolved as a vital multidisciplinary field of knowledge and practice, in consonance to increasing formalisation of management systems. In this regard, this book is timely and provides a text for teaching, research, training, and practice in Engineering Asset Management.

The book uniquely provides an innovative coalescence of multidisciplinary principles and concepts for Engineering Asset Management. Case studies are cited and discussed with regard to the management of engineered assets in several industry sectors such as agriculture, resource extraction and processing, manufacturing and production, transport, transportation and utilities in both public and private settings.

I highly recommend this book because it authentically and comprehensively addresses issues relevant for managing the vastness of engineered assets in the era of Society 5.0 concomitant with 4IR technologies and globalisation 4.0 business models. The content of the book is organised in a consistent and logical manner that should appeal to students, academics, policy makers, and practitioners alike.

Adolfo Crespo Márquez
University of Seville
Seville, Spain

Foreword by Belle R. Upadhyaya

Engineered assets such as equipment, machinery, processing lines, and cyber-physical systems are indispensable for commerce, governance, industry, and human livelihood. This timely book has innovatively brought together concepts and principles from the relevant multidisciplines and thus addresses the body of knowledge and practical concepts required for managing engineered assets. The importance of the value system, risk assessment, reliability, resilience, and sustainability imperatives, as well as managing engineered assets in accordance to legislation and standards, is reiterated throughout the book. The continuous improvements in industrial processes demand a predictive approach for managing engineered assets.

I highly recommend this book to students, academics, policy makers, and practitioners who strive to develop and demonstrate new ways to tackle the challenges of managing engineered assets in the changing industrial era of Society 5.0 and 4IR.

If you need further information, please feel free to contact me: bupadhyaya@comcast.net

Sincerely,

Belle RUpadhyaya

Belle R. Upadhyaya
Fellow, International Society of Engineering Asset
Management (ISEAM)

Fellow, International Society of Automation (ISA)

Fellow, American Nuclear Society (ANS)

Professor Emeritus
University of Tennessee
Knoxville, TN, USA

Preface

Although the practice of managing engineered assets has existed throughout human history, the formalisation of management practices has accelerated as civilisation transcends into and beyond the era of Society 5.0 concomitant with fourth industrial revolution (4IR) technologies. It is arguable that the accelerated formalisation of management practices has also spurred the emergence of Engineering Asset Management (EAM) as a multidisciplinary area of knowledge and practice. Engineered assets include all man-made things such as artwork (e.g., drawings and paintings), tools, gadgets, buildings, equipment, machines, infrastructure, industrial plant, large-scale physical facilities, and cyber-physical systems which pervade all sectors of industry and society.

This book originally combines several multidisciplinary principles and concepts in order to advance knowledge which is indispensable for managing engineered assets. The theoretical and technical representations are derived from experience, expertise, and knowledge from several conventional disciplines including philosophy, science, mathematics, engineering and technology, business management, finance, law, information and communications technology, automation and control, operations and maintenance, logistics, as well as people management.

By coalescing principles and practical concepts from the relevant multidisciplines, the book addresses the fundamental body of knowledge essential for managing engineered assets. The book serves both the academic preference for a text that facilitates teaching and research and the practitioner preference for guidelines that point towards solving practical challenges.

The philosophical, scientific, technical, and managerial knowledge areas of this book are arranged in nine chapters covering:

- Concepts ranging from investment, value, and valuation; sustainability; to uncertainty, risk, resilience, and vulnerability;
- Principles including planning, scheduling and organising, team and teaming, due diligence, and decision-making are combined with imperatives of sustainability, standards, and legislative requirements;
- Practices ranging from capital project execution, operations and maintenance, work execution, implementation of management systems coupled with data and information management, condition, risk, reliability, resilience, and

vulnerability assessments, to performance measurements and improvement opportunities;

- Practical issues such as financial and technological appraisals, technology obsolescence, the conflation and influence of 4IR technologies together with globalisation 4.0 business models.

The content of the book should be of high interest to

- Managers of engineered assets in both the public and private sectors who seek further understanding of principles and practical concepts for managing engineered assets to achieve their organisational purposes and missions;
- Scholars and students who intend to acquire, broaden, and strengthen knowledge about the principles and practical concepts involved in managing engineered assets;
- Policymakers and regulators seeking to improve policymaking, governance, assessment and evaluation frameworks regarding the management of engineered assets;
- The broader society concerned about sustainable management of engineered assets as the core activity embedded within the circular economy model.

I must thank all persons that I have fortunately encountered throughout my academic and industry careers. A great number of you have helped me write this book through your reviews of my academic papers and your interactions with me at conferences, seminars, workshops, and during my consulting and industry work assignments. Thank you to all the fellows and members of the International Society of Engineering Asset Management (ISEAM) and also the fellows and members of the International Association for the Management of Technology for the many fruitful discussions over the years. My students and mentees, as well as my colleagues past and present all deserve mention for encouraging, persuading, and generally supporting my efforts over a considerable number of years. I also wish to thank future students, mentees, and colleagues in advance as I envisage that you will be motivating my future endeavours. Thank you to the staff at Springer for making the book publication process smooth and rewarding. All errors in the text are mine.

Pretoria, South Africa   Joe E. Amadi-Echendu
June 2021

Contents

About the Author

Joe E. Amadi-Echendu earned BS and MS degrees in Electrical and Electronic Engineering at the University of Wyoming and DPhil in Control Engineering at the University of Sussex. He has served as a teaching assistant at Wyoming, a research fellow at Sussex, a lecturer at the University of Port Harcourt and Rivers State University of Science and Technology, and a senior lecturer at the University of Greenwich. He currently serves as a tenure track professor of Engineering and Technology Management at the University of Pretoria. He founded an Institute of Engineering, Technology, and Innovation Management at the University of Port Harcourt. He commenced his career as an apprentice maintenance technician at a tyre manufacturing facility, and his expertise is underpinned by his doctoral research in digital signal processing, condition monitoring, diagnostics and prognostics of engineered assets. His extensive experience includes 12 years of work focussed on managing engineered assets in resource extraction and minerals processing industry, as well as 30+ years in academia, including consultancy assignments in manufacturing and utility sectors. His teaching, research, and consulting work experiences have resulted in the publication of over 150 articles and 6 co-edited books in the multidisciplinary fields of Engineering Asset Management and Technology Commercialisation. He is the founding editor-in-chief of Springer's Engineering Asset Management Review Series. He is a registered professional engineer in the UK and South Africa, and a member of a number of international institutions. He is a founding fellow and currently serves as a chairperson of the International Society of Engineering Asset Management, as well as a treasurer of the board of the International Association for the Management of Technology.

Definitions and Scope

1

Abstract

This chapter provides a brief articulation of the motivation for managing engineered assets that facilitate and support all facets of human endeavour, livelihood, and industry. The articulation is based on a brief but concise discourse on some of the historical antecedents to the wide ranging *cross*-disciplinary, *inter*-disciplinary, *trans*-disciplinary, and *multi*disciplinary human endeavour of managing engineered assets. The chapter includes an overview of the grand challenges for engineering in the sustainability driven era of *Society 5.0* powered by combinations of 1st, 2nd, 3rd, and 4th industrial revolution (4IR) technologies. This introductory discourse concludes with definitions for 'asset' and 'asset management'.

1.1 EAM in the Era of Society 5.0 and 4IR

Throughout history,[1] assets remain indispensable and invaluable to human existence, livelihood and industry. Early human beings certainly knew how to create, utilise and discard their tools at the end of useful life—archaeologists excavating ancient sites continue to uncover some of the tools. The earliest tools[2] were created from what is available in nature, and provided the means for the hunter-gatherer livelihood and nomadic way of life of early humans. Archaeological records and

---

[1]Britannica History of Technology in the Ancient World. https://www.britannica.com/technology/history-of-technology/Technology-in-the-ancient-world.
[2]Groeneveld, E. (2016). Stone age tools. In *Ancient history encyclopedia*. https://www.ancient.eu/article/998/.

© The Author(s), under exclusive license to Springer Nature Switzerland AG 2021
J. E. Amadi-Echendu, *Managing Engineered Assets*,
https://doi.org/10.1007/978-3-030-76051-9_1

on-going discoveries (e.g., in Egypt, Greece, and many other parts of the world) show that humans engineered assets that crucially facilitated governance, commerce and industry in earlier civilisations.

As civilisations evolved, modern humans created and utilized assets that provided the means to harness water and steam to power the 1st industrial revolution of mechanized production. The scope, range, and variety of engineered assets continued to increase as humans harnessed electricity to power the 2nd industrial revolution. The impetus for the 3rd industrial revolution is credited to information and communications technologies. Today, we still engineer assets from nature's providence, and the scope, range, and variety of assets continue to evolve as human communities, economies, and societies transcend to the era of *Society 5.0* powered by the cumulative technologies of the 1st, 2nd, 3rd, and 4th industrial revolutions. As civilization hurtles towards trajectories that will manifest in the future, engineered assets will remain indispensable to human livelihood and industry, and the management of engineered assets towards our seemingly insatiable desire for instant gratification will always remain a challenge.

In acknowledging that the grand challenges are consistent with the sustainability paradigm, the discourse in this book is contextualised in the era of *Society 5.0* powered by the cumulative technologies of the industrial revolutions. Table 1.1 provides a summary of the characterisation of the era of *Society 5.0* subsuming that the management of engineered assets or 'asset management' transcends all eras of human civilization.

1.2   EAM is a Grand Challenge for Sustainable Development

"Products, [services] and processes that enhance the joy of living remain a top priority of engineering innovation … and the world's cadre of engineers will continue to seek ways to put knowledge into practice to meet the grand challenges[3]" depicted in Fig. 1.1.

It is widely acknowledged that there is a critical need to inspire and educate future generation of engineers capable of confronting and addressing the grand challenges for engineering. In this regard, this book is about the management of engineered assets, that is, man-made artefacts or things that facilitate all aspects of human endeavour, livelihood, industry, and enhance the human way of life.

Engineered assets, that is, man-made artefacts or things range from artwork (e.g., drawings and paintings), personal gadgets and tools to buildings, equipment, machines, industrial plant, physical facilities and infrastructure, and, as well as the conflation of cyber-physical systems (CPS) that are facilitating the fusing of our biological, digital, and physical worlds. In fact, it is mind boggling to recount the range of man-made things that we utilise and rely on for our daily existence and

---

[3]Grand Challenges for Engineering. www.engineeringgrandchallenges.org.

Table 1.1 A mapping of eras of human civilisation (see also Fukuyama (2018); Salgues (2018); Schwab (2016); Bennet and Lemoine (2014))[Fukuyama, M. (2018). Society 5.0: Aiming for a new human-centered society. Japan spotlight, July/August 2018.; Salgues, B. (2018). *Society 5.0: Industry of the future, technologies, methods and tools.* ISTE Ltd and Wiley, 2018. ISBN 978-1-78630-301-1.; Schwab, K. (2016). *The fourth industrial revolution.* Crown Business.; N. Bennet, N., & Lemoine, G. J. (2014). *What VUCA really means for you.* Harvard Business Review, Jan–Feb 2014.]

| | | Era | | |
Industrial Revolution	Characteristic	Paradigm	Society	Characteristic
-		Ancient or Pre-Modern	1.0	▪ Gathering & hunting ▪ Survival of early societies
1.0	• Steam power • Mechanization • Knowledge formulation	Modern	2.0	▪ Agricultural ▪ Rapid formalisation of societies ▪ Fragmented industrialisation
2.0	• Electricity • Mass production • Knowledge evolution		3.0	▪ Massive industrialisation ▪ Effectiveness ▪ Geopolitics
3.0	• Information technology • Automation • Knowledge distribution		4.0	▪ Globalized industries, institutions & markets ▪ Global connectivity ▪ Global dependency
4.0	• VUCA • Cyber physical systems • Knowledge mutation	Sustainability	5.0	▪ VUCA (volatile, uncertain, complex, ambiguous) ▪ Information about everything ▪ Instant gratification

Fig. 1.1 Grand challenges for engineering. *Source* www.engineeringgrandchallenges.org

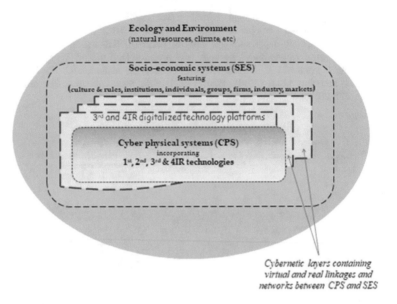

Fig. 1.2 Engineered assets as CPS are encapsulated within SES, the environment and ecology

livelihood! Just imagine that we do not have mobile phones, buildings to live and work, electricity, water from taps, roads and motorised vehicles, airports and airplanes, facilities to manufacture and process the plethora of things that we deploy and utilise in daily living, et cetera.

The grand challenges for engineering more or less highlight that engineered assets constitute CPS that underpin governance, commerce, and industry. Engineered assets are not only embedded within a socio-economic world but more significantly, engineered assets and socio-economic business activities are both encapsulated within the environment and ecology. Thus, as illustrated in Fig. 1.2, engineered assets and socio-economic activities that serve humanity are inextricably interwoven within the environment and ecology. This intrinsic and interweaving reciprocity between engineered assets, socio-economic human endeavour, the environment and ecology means that engineered assets must be managed within the precepts of the sustainability paradigm.

Curiously, it can be surmised that the increasing formalisation of management practices more or less manifest the operationalization of the sustainability paradigm. Therefore, the management of engineered assets or 'asset management' constitutes a primordial basis for operationalizing sustainability towards achieving the targets of the sustainable development goals.[4] In a sense, the emergence of 'asset management' or the management of engineered assets as a field of knowledge and practice represents the world's cadre of engineers and humanity's attempts to tackle the grand engineering challenges and sustainability imperatives. It is in this regard that some of the more recent antecedents are surmised as follows.

---

[4]UN Sustainable Development Goals (2015). https://sdgs.un.org/goals.

1.3   EAM Historical Antecedents

1.3.1   Terotechnology

Whereas asset management has been happening throughout human history, however, a more recent motivation for managing engineered assets emanates from increasing concerns about ageing infrastructure especially in developed countries. The consequent trend since the 1970s has been to increasingly formalize the management of public infrastructure assets. Notably, circa 1975, the United Kingdom Committee for Terotechnology defined terotechnology as "a combination of management, financial, engineering and other practices applied to physical assets in pursuit of economic life cycle costs". According to the Committee

> the nature of the maintenance activity was determined by the manner in which plant and equipment was designed, selected, installed, commissioned, operated, removed and replaced. Major benefits could come to embrace all these areas, and because no suitable word existed to describe such a multidisciplinary concept, the name 'terotechnology' (based on the Greek work 'terein' - to guard or look after) was adopted".

Terotechnology is a multidisciplinary concept and the associated standard[5] states that the aim is to achieve the 'best possible value for money for a user from the procurement and subsequent deployment of a physical asset'. After all, engineered assets represent the physical form of technology, and because the engineering and technology disciplines constitute *diadelphous* and *dicephalous* dualism, it is not surprising that the term 'terotechnology' became a precursor to the terminology of 'physical asset management'.

The trend towards increased formalization of the management of engineered assets has been indirectly spurred on by the business re-engineering[6] frenzy of the 1980's, which entailed:

i. fundamental rethinking of the way work gets done so as improve productivity by concomitantly reducing cycle times;
ii. structural reorganisation of vertical functional hierarchies into horizontal and matrix cross-functional teams;
iii. exploitation of information and communication technologies to establish data, information, and performance measurement systems for improved decision making;
iv. greater appreciation for uncertainty, plus improved understanding of risks coupled with emphasis on providing value to stakeholders.

---

[5]BS 3843. (1992). Terotechnology (the economic management of assets). British Standards Institution.
[6]Rigby, D. (1993). "The secret history of process reengineering. *Planning Review, 21*(2), 24–27.

1.3.2 Safety and Legislative Antecedents

(a) *Safety*

Among others, one remarkable risk event that can be linked to increased fervour to formalise the management of physical[7] assets was the 6 July 1988 Piper Alpha disaster that resulted in 167 deaths and estimated property damage of £1.7 billion. The Cullen[8] report and recommendations still provide lessons on the management of large scale assets operated in very hazardous environments. The frenzy and turmoil of the Piper Alpha disaster increased awareness on statutory regulations, albeit that the focus was mostly on safety. A more recent event is the 2010 Deepwater Horizon[9] oil spill that killed 11 people, causing extensive impact on the environment and ecology, and resulting in extensive legal actions.

Sadly, on 4 August 2020, one of the biggest non-nuclear explosions that devastated the area of the port of Beirut resulted in more than 180 human fatalities with over 6000 persons injured. According to a report,[10] the blast caused widespread destruction, damaging "about half the buildings in Beirut, displacing more than 250,000 people; and silos holding 85% of the country's wheat stores were either destroyed or were so badly damaged that the grain was no longer edible." Tantalizingly, the news generally include accounts of man-made accidents and natural disasters that result in fatalities, destruction of engineered assets, and inducing complex effects on the environment and ecology.

(b) *Legislation*

Legislative adoption of accrual accounting in the public sector also required improved formalisation of asset management to bring about long term change.[11] Global accounting standards (e.g., international financial reporting standards (IFRS), generally accepted accounting principles (GAAP), generally recognised accounting practices (GRAP), and the triple bottom line (TBL)[12] tenets) require that assets are valued accurately, and resources are appropriately leveraged to ensure

---

[7]Here, physical assets include natural resources like minerals and engineered assets like infrastructure.

[8]Cullen, W. D. (1990). *The public enquiry into the Piper Alpha disaster*. Department of Energy, HMSO, London.

[9]Pallardy, R. (2010). *Deepwater horizon oil spill*. https://www.britannica.com/event/Deepwater-Horizon-oil-spill.

[10]Ramzy, A., & Peltier, E. (2020). *What we know and don't know about the Beirut explosions*. The New York Times, 5 August 2020. https://www.nytimes.com/2020/08/05/world/middleeast/beirut-explosion-what-happened.html.

[11]Harris, R. (2010). *Public sector asset management: A brief history*.

[12]Hammer, J., & Pivo, G. (2016). The triple bottom line and sustainable economic development theory and practice. *Economic Development Quarterly*, 1–12. https://doi.org/10.1177/0891242416674808.

that capital is efficiently and effectively employed. Whilst managers of private sector assets purportedly grapple with the requirements of IFRS and GAAP, accrual accounting further demands that managers of public sector assets contend with fiscal prudence[13] in budgetary execution, efficiency in economic resource allocation, social equity in service delivery, and adherence to ecological and environmental sustainability imperatives.[14]

The trend for better management of public infrastructure assets is exemplified by the antecedents to the World Bank's "vision of sustainable cities that are livable, competitive, well-governed and managed, and bankable."[15] Many municipal, regional and state governments continue to aim and pursue this ideal. Literature cited in the Worley International Report[16] surmises approaches adopted by many municipal, regional and state governments towards managing public infrastructure assets. The report states that, to achieve 'best practice' in asset management, organizations must be able to demonstrate:

i. knowledge of levels of service required by customers;
ii. ability to predict future demands for service;
iii. knowledge of ownership of existing assets;
iv. knowledge of physical condition of assets;
v. knowledge of asset performance and reliability;
vi. knowledge of asset utilization and capacity;
vii. ability to predict the failure modes and estimated time of failure for assets;
viii. ability to analyse alternative treatment options;
ix. ability to rank works based on economic analysis;
x. ability to prioritize works to suit the available budget;
xi. ability to develop and revise strategic objectives for each asset;
xii. ability to optimize operations and maintenance activities.

1.3.3 Other Antecedents

The ability to predict the failure modes and estimated time of failure for assets particularly links to the much acclaimed work of Nolan and Heap[17] on reliability-centered maintenance (RCM). The objective of the RCM discipline is to

---

[13]Grubišić, M., Nušinović, M., & Roje, G. (2009). Towards efficient public sector asset management. *Financial Theory and Practice*, *33*(3) 329–362.

[14]Amadi-Echendu, J. E. (2013). Assessment of engineering asset management in the public sector. In *Proceedings of 8th WCEAM and 3rd ICUMAS* (pp. 1151–1156), Hong Kong Oct 31–Nov 1 2013. Springer e-book. ISBN 978-3-319-09507-3.

[15]Cities in transition: World Bank Urban and Local Government Strategy. (2000). ISBN 0-8213-4591-5.

[16]Urquhart, T., & Busch, W. (2000). *Strategic municipal asset management*. Worley International Ltd. The World Bank.

[17]Nowlan, F. S., & Heap, H. F. (1978). *Reliability-centered maintenance*. Dolby Access Press.

develop necessary scheduled maintenance activities towards the realization of the inherent reliability and avoidance of failure modes of engineered artefacts (e.g., equipment). The RCM philosophy supposedly considers consequences of equipment failure on safety, operations and economics, hence the goal of RCM is to perform scheduled tasks "necessary to protect safety and operating reliability, ... and to do so at minimum cost."

The term 'asset management' also arises from the perspective of maintenance, repair, and operations (MRO). With the MRO convention, the focus was on the supply of consumables like lubricants for greasing, plus tools and spares for repair of equipment, especially compressors, pumps, and valves. This MRO antecedent provides the links between supply chain management, production/manufacturing resource planning (MRP), and maintenance management. A closely related antecedent consequent upon the knowledge management bias is product lifecycle management (PLM) where manufacturers re-use (cycle) intellectual capital in the form of product related data, knowledge and processes to fast track development and launch of new products.[18] It is arguable that the term 'life cycle' gained prominence in terms of PLM and still prevails.

A parallel antecedent emanates from the rapid evolution and application of information and communication technologies (ICT) for the automation of business processes and activities. The 'explosion' of enterprise resource planning (ERP) systems also gave rise to a plethora of computerized maintenance management systems (CMMS). The RCM discipline and CMMS are widely adopted especially for the maintenance of equipment and machinery. Terminologies like 'run-to-failure', corrective-based, preventative-based, predictive-based, proactive, opportunity-based, and condition-based maintenance have become widespread and sometimes perplexing. Since the 2000s, ERP, CMMS, RCM software, and several ICT acronyms have been conflated into so called enterprise asset management systems. Subsequently, emphasis shifted to the implementation of ICT systems, culminating in the current ISO 5500x series of standards which are primarily concerned with management systems for asset management.

The prominence given to ICT-based management systems has become alluringly epitomized by the ISO 5500x standards concomitant with a frenzy to rebrand maintenance management.[19] Interestingly, frenetic new terms like 'asset care' and 'asset health monitoring' tend to obfuscate and vitiate the scope of asset management.[20] Although the ISO 5500x standards provide conformity with regard to asset management systems, however, as depicted later in this book, the multidisciplinary nature of asset management means that there is a myriad of standards from the many disciplines that constitute the body of knowledge for theory and practice of asset management.

---

[18]Javvadi, L. (2015). *Product life cycle management: An introduction.* https://www.researchgate.net/publication/285769844.

[19]ISO/TC 251 WG4 Managing assets in the context of asset management, May 2017.

[20]Amadi-Echendu, J. E. (2017). Developments in the management of engineered assets: before and beyond ISO5500x. Keynote address 4—IEOM Conference, Rabat, Morocco, 11–13 April 2017.

As mentioned earlier, the scope, range, and variety of engineered assets that support human existence and livelihood is extremely wide. Furthermore, human endeavour encompasses many disciplines, so it not surprising that there will be a number of definitions for 'asset' and 'asset management'. Some of the pertinent definitions are discussed as follows.

1.4   Asset Definitions

1.4.1   What is an Asset?

The simplest definitions of an asset are that it is "a useful or valuable thing, or person"; "an item, thing or entity that has potential or actual value…" In the context of investments, "an asset is a resource with economic value that an individual, corporation or country owns or controls with the expectation that it will provide a future benefit." In financial accounting context, an asset is a tangible or intangible resource that can be owned and/or controlled (utilized) to produce value. The IASB[21] Framework defines an asset as "a resource controlled by the entity as a result of past events and from which future economic benefits are expected to flow to the entity." In legal terms, an asset is a useful or valuable property which might be available for the repayment or liquidation of debt. For the purposes of this book, the emphasis will be on managing a useful or valuable thing, thus, we shall restate that an asset is a useful **and** valuable thing. Furthermore, we define an engineered asset as a man-made artefact which provides the means for the realization of value.

These definitions highlight a number of pertinent keywords for the management of assets. The first keyword is *value*—not only is value imputed to an asset *ab initio*, but *de facto*, an asset provides the means for the realization of value (*benefits*). Tantalizingly, benefits can be described in both quantitative and qualitative terms. The fact that there is accounting, engineering, finance, and legal definitions of an asset suggests that the value of an asset may be viewed differently by the disciplines. This means that determining the value of an asset may not be trivial. In fact, at a given point in time, an asset manager would be obliged to determine a value for an asset, i.e., provide a *valuation* of an asset.

The second and subtle keyword is *valuation*—re: economic value, repayment or liquidation of debt. While value is virtuously inherent in an asset, however, valuation depends on the viewpoint of the valuer. Valuation forms part and parcel of managing an asset, thus, an asset manager should be able to establish a valuation of the asset.

The third keyword is *useful*—implying that, if an asset is not useful, it destroys the intrinsic value of the asset. Thus, an asset manager should be able to determine the utilization of an asset. Utilization may be stipulated either in terms of capacity, or time, or both. For example, utilization may be a compounded measurement of the

---

[21]International Accounting Standards Board.

number of passenger seats occupied (capacity) during a scheduled flight (duty cycle) of a commercial aircraft.

The fourth keyword is *property* or *ownership*—meaning that an entity must have full legal rights and accountability for an asset, that is, capability to utilise the asset and to appropriate the benefits accruing thereof. A closely related keyword is *control*—meaning that an entity that may or may not possess full legal rights must be held responsible and accountable for an asset. Interestingly, the asset manager may control the asset at the behest of the asset owner, that is having custodian, steward, and usufruct rights and privileges.

The fifth keyword is *resource*—a synonym for an asset. The asset manager should treat the asset as a resource, that is, as a means for the realization of benefits which encompass both quantitative and qualitative value metrics. A conundrum arises in the sense that an asset not only possesses intrinsic value but also, the asset also provides the means for the realisation of value! Extrapolating the terotechnology tenet, these keywords re-iterate that the capacity for asset management encompasses multidisciplinary roles, responsibilities and accountabilities. Before discussing this very important issue further in later sections of the book, it is necessary to explain categories and types of assets.

1.4.2 Asset Categories and Types

The categorization or classification of assets depends on perspective. In terms of the IASB Framework (see Fig. 1.3), assets are generally classified in three ways, based on:

- convertibility-based on how easy it is to convert a physical asset into cash;
- physical existence—tangible or intangible form;
- usage-based on what an asset is utilized for, or how it is deployed.

This book is primarily concerned with the management of engineered assets and covers the respective hierarchical[22]:

i. assembly, aggregation, combination, conglomeration, or integration of man-made artifacts into buildings, equipment, machinery, infrastructural facilities, industrial plant and cyber physical systems, and
ii. resolution of man-made systems into components, parts, assemblies and subsystems.

As depicted in Fig. 1.4, it is worth pointing out that physical assets refer to both natural and engineered assets. Where necessary and applicable, the discourse will also include financial assets (such as stocks, bonds), natural assets (such as water,

---

[22]International Atomic Energy Agency. (2021). Asset Management for Sustainable Nuclear Power Plant Operation. IAEA Nuclear Energy Series No. NR-T-3.33. www.iaea.org/publications.

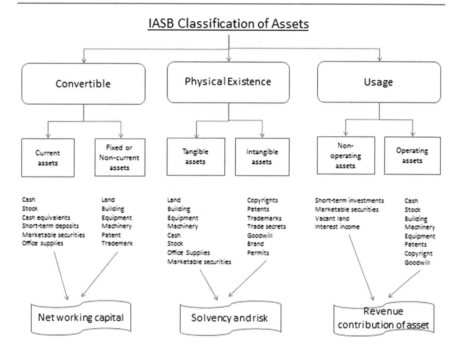

Fig. 1.3 IASB classification of assets

minerals), people, technology and intangible assets. There are other categorizations of physical assets that engender specialised management focus, some of these include:

- Facilities and infrastructure—i.e., airports, bridges, roads, hospitals, residential and non-residential buildings, stadiums, etc.; leading to the specialised area of facilities and infrastructure management[23] where the integration of people, place and process with the built environment is managed to improve quality of life and productivity of human endeavour.
- Fleets of assets—e.g., airplanes and land-based automobiles, leading to the specialised area of fleet management, particularly in the transportation industry.
- Industrial plant—featuring equipment, machines, and facilities that enable raw materials extraction, processing, refining, manufacturing, reticulation, distribution, recycling, etc.
- Farms, gardens, parks and nature reserves where the natural environment is typically cultivated by human societies for the purposes of agriculture, adventure and pleasure.

---

[23]ISO 41011:2017 Facility Management—Vocabulary.

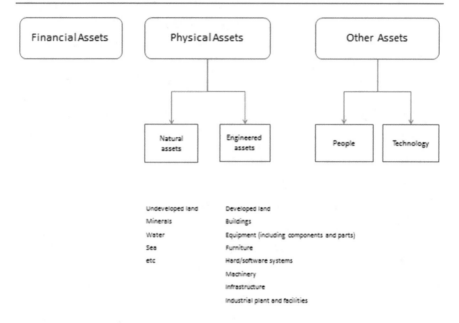

Fig. 1.4 Types of assets

It is worth mentioning here that there are other important categorizations or classifications physical of assets which also give rise to specialized management emphasis, viz:

- public sector assets—where traditionally, the public sector and government agencies own and control the assets for the primary purpose of service delivery;
- private sector assets—where the private sector industry conventionally owns and controls the assets for the primary purpose of commercialisation, financial gain and the creation of triple-bottom-line value;
- public-private assets—this increasingly popular classification leads to various partnership arrangements between the public and the private sectors.

1.4.3 Discrete Assets, Asset Systems, Cyber Physical and Socio-Technological Systems

It is not uncommon to view a personal gadget (e.g., mobile phone), a tool, or even a personal vehicle as a discrete asset. Even so, given the era of Society 5.0 and fourth industrial revolution (4IR) technologies, the reality is that there are very few discrete assets that are not interconnected and dependent on other assets. In fact, the evolution of cyber-physical systems (CPS) confirms and buttresses the fact that the pragmatic view should be in terms of systems of assets or asset systems. After all, a CPS is an engineered system that is "built from, and depends upon the seamless integration of computational algorithms with physical components". A CPS is an asset system characterized by the conflation and embedding of computational algorithms and electronic communication capabilities in the constituting physical artefacts, components, objects, structures and subsystems. The embedded computing algorithms and communication capabilities thus enable the operations, coordination, control, maintenance, monitoring and support, that is, the overall management of both relatively 'simple assets' as well as complex and sophisticated machinery and interdependent infrastructure systems.

CPSs are primordial to all human endeavor and industry and thus constitute infrastructure for the provisioning of education, energy and water, health, security, and transportation, etc. That is, CPSs are inextricably interwoven with socio-economic business activities in the service of humanity. In the era of Society 5.0, the fusion of CPS with socio-economic business activities is facilitated by the digitalized platforms of 4IR technologies. In fact, socio-technological systems are constituted by digitalized technology fusion of CPS with socio-economic business activities as encapsulated within the world of ecology and the environment (see Fig. 1.5). Therefore, throughout this book, the word 'asset' or 'assets' may denote discrete, simple or complex and sophisticated system or systems of assets.

- cyber physical systems (**CPS**) are embedded within socio-economic systems (**SES**) and facilitate socio-economic business activities

- CPS and SES are inextricably interwoven within ecology and the environment (**E&E**)

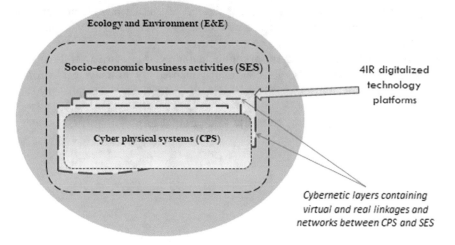

Fig. 1.5 Engineered assets within socio-technological systems

1.5 Definitions of Asset Management

As may be evident from the discourse so far, the words 'asset' and 'management' respectively have broad and ubiquitous meanings, so, there are several definitions for asset management. For brevity, some of the definitions are re-stated from various references as follows.

Asset management

- "is the combination of management, financial, economic, engineering, and other practices applied to physical assets with the objective of providing the best value level of service for the costs involved."
- "… refers to systematic approach to the governance and realization of value from assets."
- is "a process that guides the gaining of assets, along with their use and disposal in order to make the most of the assets and their potential throughout the life of the assets."
- "… is the process of ensuring that assets are maintained, accounted for, and put to their highest and best use."
- "… is the way in which the acquisition, use and disposal of the assets … are managed in order to maximize any profit they generate."
- "… is a systematic process of developing, operating, maintaining, upgrading, and disposing of assets in the most cost-effective manner (including all costs, risks and performance attributes)."
- "is the practice of managing the entire life cycle (design, construction, commissioning, operating, maintaining, repairing, modifying, replacing and decommissioning/disposal) of physical [engineered] assets."
- "combines science, engineering, and ergonomic principles with sound business practice and technology, taking cognizance of sustainability constraints, to facilitate informed decisions that are necessary for an asset to provide the value intended."
- "involves acquisition, ownership, control and use of a physical asset throughout its life stages to satisfy the constraints imposed by business performance, economy, ergonomics, technical integrity, regulatory compliance, and sustainability."
- "is coordinated activity … to realise value from assets."[24]

---

[24]ISO 55000:2014. Asset management—Overview, principles and terminology.

In this book, we shall define engineering asset management[25] to include all activities which ensure that an engineered asset provides the means for the realization of value.[26] It is clear from the definitions of asset and asset management that the scope of asset management is, indeed, very wide. So it is not surprising that asset management confounds many organisations as it is challenging to compartmentalize into conventional or traditional organisational functions of engineering, facilities, finance, maintenance, operations, production, marketing, technology, etc. In practice, every entity will define asset management as best suits the entity's purpose.

1.6 Summary of the Chapter

In this chapter, some historical antecedents have been highlighted in order to contextualize the more recent precursors to the subject matter of engineering asset management. A number of definitions that are crucial to managing engineered assets have been discussed. The various definitions confirm the multidisciplinary nature of asset management and point to a number of principles and practical concepts that are crucial for engineering management of assets. These principles and concepts constitute the core of this book, and are discussed in further detail in the following chapters.

1.7 Exercises

1. Imagine that you have been recruited as the 'asset manager' for a particular organisation in a particular public or private sector industry (e.g., agriculture and agro-allied, education and research, health, hospitality, manufacturing, mining and minerals processing, security, transportation, utilities, etc.). How would you define engineering asset management for the organisation?
2. Based on the discourse in this chapter, state at least six things that an asset manager should do.

---

[25]Hodkiewicz, M. R., & Pascual R. (2006). Education in engineering asset management—Current trends and challenges. In *International Physical Asset Management Conference*, Tehran, 28th–31st 2006.
[26]Amadi-Echendu et al. (2010) What is Engineering Asset Management: Chapter in Engineering Asset Management Review, January 2010. https://doi.org/10.1007/978-1-84996-178-3_1.

References and Additional Reading

Amadi-Echendu, J. E. (2013). Assessment of engineering asset management in the public sector. In *Proceedings of 8th WCEAM and 3rd ICUMAS* (pp. 1151–1156), Hong Kong Oct 31–Nov 1 2013. Springer e-book. ISBN 978-3-319-09507-3.

Amadi-Echendu, J. E. (2017). Developments in the management of engineered assets: before and beyond ISO5500x. Keynote address 4—IEOM Conference, Rabat, Morocco, 11–13 April 2017.

Amadi-Echendu, J. E., Willet, R. J., Brown, A. B., Hope, T., Lee, J., Mathew, J., Vyas, N., & Bo-Suk, Y. (2010). What is engineering asset management: Chapter in engineering asset management review (Vol. 1). https://doi.org/10.1007/978-1-84996-178-3_1.

Bennet, N., & Lemoine, G. J. (2014). *What VUCA really means for you.* Harvard Business Review, Jan–Feb 2014.

Britannica History of Technology in the Ancient World. https://www.britannica.com/technology/history-of-technology/Technology-in-the-ancient-world.

BS 3843. (1992). Terotechnology (the economic management of assets). British Standards Institution.

Cities in transition: World Bank Urban and Local Government Strategy. (2000). ISBN 0-8213-4591-5.

Cullen, W. D. (1990). *The public enquiry into the Piper Alpha disaster.* Department of Energy, HMSO, London.

Fukuyama, M. (2018). Society 5.0: Aiming for a new human-centered society. Japan Spotlight, July/August 2018.

Groeneveld, E. (2016). Stone age tools. In *Ancient history encyclopedia.* https://www.ancient.eu/article/998/.

Grubišić, M., Nušinović, M., & Roje, G. (2009). Towards efficient public sector asset management. *Financial Theory and Practice, 33*(3), 329–362.

Hammer, J., & Pivo, G. (2016). The triple bottom line and sustainable economic development theory and practice. *Economic Development Quarterly*, 1–12. https://doi.org/10.1177/0891242416674808.

Harris, R. (2010). *Public sector asset management: A brief history and outlook.* Ramidus Consulting Limited. http://www.ramidus.co.uk/papers/publicsectorassetman.pdf.

Hodkiewicz, M. R., & Pascual R. (2006). Education in engineering asset management—Current trends and challenges. In *International Physical Asset Management Conference*, Tehran, 28th–31st 2006.

International Atomic Energy Agency. (2021). Asset Management for Sustainable Nuclear Power Plant Operation. IAEA Nuclear Energy Series No. NR-T-3.33. www.iaea.org/publications.

ISO 55000:2014. Asset management—Overview, principles and terminology.

ISO/TC 251 WG4 Managing assets in the context of asset management, May 2017.

Javvadi, L. (2015). Product life cycle management: An introduction. https://www.researchgate.net/publication/285769844.

National Academy of Engineering. (2008). Grand Challenges for Engineering. www.engineeringgrandchallenges.org.

Nowlan, F. S., & Heap, H. F. (1978). *Reliability-centered maintenance.* Dolby Access Press.

Pallardy, R. (2010). *Deepwater horizon oil spill.* https://www.britannica.com/event/Deepwater-Horizon-oil-spill.

Ramzy, A., & Peltier, E. (2020). *What we know and don't know about the Beirut explosions.* The New York Times, 5 August 2020. https://www.nytimes.com/2020/08/05/world/middleeast/beirut-explosion-what-happened.html.

Rigby, D. (1993). The secret history of process reengineering. *Planning Review, 21*(2), 24–27.

Salgues, B. (2018). *Society 5.0: Industry of the future, technologies, methods and tools.* ISTE Ltd and Wiley. ISBN 978-1-78630-301-1.

Schwab, K. (2016). *The fourth industrial revolution.* Crown Business.

UN Sustainable Development Goals. (2015). https://sdgs.un.org/goals.

Urquhart, T., & Busch, W. (2000). *Strategic municipal asset management*. The World Bank: Worley International Ltd.

Value and Sustainability

2

Abstract

As a matter of emphasis, it pertinent to reiterate that engineered assets range from artwork, personal gadgets and tools to buildings, equipment, machines, infrastructure, industrial plant, as well as large scale physical facilities, and the cyber-physical conflation of systems that represent the fusing of biochemical, digital, and physical worlds. The discourse in this chapter will highlight the value ethos and the sustainability imperative as fundamental principles for the management of engineered assets. Starting with value as an ethos, the discourse includes investment and valuation concepts as well as a concise discussion on the sustainability imperative. These principles are coalesced from a number of the traditional disciplines involved in the management of engineered assets.

2.1 Value

From a philosophical viewpoint, the terms value,[1] investment, and valuation are completely intertwined. After all, the purpose of an investment is to obtain value, and value is established through the process of valuation. An understanding of the value ethos together with the associated concepts of investment and valuation is primordial to making decisions about the management of engineered assets.

---

[1]"Values are the basis of personal and collective judgments about what is important in life - influenced by culture, religion, and laws. They are factors in our decision-making on social, environmental, and political matters, and on the best uses of our time, money, and natural resources".

Value[2] is an ethos. It is "the regard that something is held to deserve; the importance, worth, or usefulness of something." It usually involves objective and subjective appraisal, assessment, estimation, evaluation, or judgment of advantage, benefit, desirability, effectiveness, practicality, quality or utility of something. The reality and conundrum is that the objective and subjective aspects are not mutually exclusive, even though the practical tendency is to mute subjectivity by stating value in quantitative terms. For an object like a family photograph or a souvenir, the value of the asset is sentimental in the sense that there is a sharp contrast between the quantitative and qualitative aspects of value. The quantitative cost of acquiring a souvenir may not be readily compared to its memorial significance or qualitative benefit!

Subjectivity derives mainly from both the epistemological, ontological and psychological[3] dimensions wherein value is interpreted in terms of behavioural strength of desire. Such measurement of value is not necessarily equivalent to a rational evaluation of the utility of an asset. Subjectivity implies that the determination of the value of an asset depends on how each stakeholder perceives the importance, worth, or usefulness of the asset. Take for example, an aeroplane—the operator has a perception of the value and will determine the value of the aeroplane based on this perception. The passenger has a perception of value which is often based on a comparison between the quantitative and qualitative aspects—the airfare versus comfort, safety, timeliness, etc. The regulator of aviation also has a perception of the value of the aeroplane. The tiers of manufacturers and suppliers of the airplane components, spare parts, equipment, subsystems and systems, as well as other supply chain vendors have varied views of the value of the aircraft. Furthermore, the insurer of the aeroplane establishes a value, albeit, primarily from the viewpoint of risk, and sometimes uncertainty. The point here is not only that the stakeholders are many and varied, but also, that each stakeholder will describe and/or determine the value of the aeroplane based on stakeholder's own objective and subjective perceptions.

Figure 2.1 is an illustration of the varied attributes of the value of an asset as may be perceived by various stakeholder groups. The entity managing the asset must not only achieve its own value specifications in terms of its business objectives, but also, it must comply with the value specifications of other groups of stakeholders such as the investor/shareholder, customer/consumer, and the community/regulator. The investor and shareholder perception of value is psychologically anchored around some form of return on investment.

The customer primarily perceives the value of an asset from the service delivered but determines the value provided by the asset in terms of price and quality[4] of service. The community's perception of the value of an asset often manifests in

---

[2]Pauls, R. (1990). Concepts of value: A multidisciplinary clarification. ISBN 1-86931-036-5.

[3]Kilmann, R. H. (1981). Toward a unique/useful concept of values for interpersonal behaviour: A critical review of the literature on value. *Psychological Reports, 48*, 939–959.

[4]Sidorchuk, R. (2015). The concept of value in the theory of marketing. *Asian Social Science, 11* (9). ISSN 1911-2025.

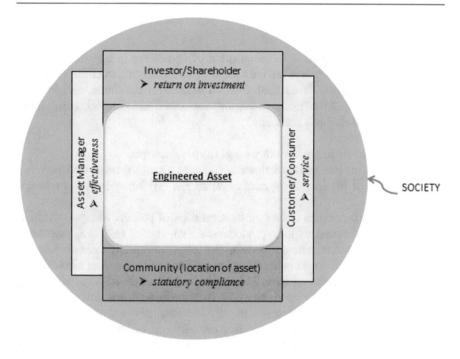

Fig. 2.1 Stakeholder perceptions of the value of an engineered asset

demands that the asset complies with society's norms which are typically stated as statutory requirements, e.g., compliance to environmental legislation and social responsibility regulations. Therefore, as illustrated in Fig. 2.1, an asset is 'surrounded' by a profile of values depicting the various stakeholders' perceptions, and each stakeholder expects that the asset will be managed to comply with the respective stakeholder's value preferences and specifications. In relative terms, the assessment of 'comfort' by a passenger in an aeroplane is likely to be more stringent than that of the airline operator because 'comfort' and convenience may be reasons for the passenger to use the aeroplane for the travel.

The existence of an asset represents an implicit fundamental value. This means that an asset is imputed with intrinsic value *ab initio*. From the asset management viewpoint, the value and usefulness (utility, utilization) of an asset are inextricably interwoven and correlated. The implication is that the value of an asset cannot be determined independently from how the asset is utilised. The practical suggestion is to specify what is desired, that is, the purpose of investing in an asset, then to manage the asset to achieve the objective of the investment.

2.2   Investment

In general, an investment is the acquisition of something (a resource) with the expectation that the thing will not only preserve its value but also create value, i.e., provide benefits. Every investment decision deals with the future, and involves alternative choices. Investment appraisals[5] are about assessing expected future benefits based on methods such as

- estimations of (i) future cash flows and (ii) pay-back period;
- analyses of (iii) profitability index and (iv) net-present value; as well as
- calculations of (v) accounting rates-of-return and (vi) internal rates of return.

An investment conjures or creates an impression of productive capacity, so, an asset must have the capacity to be productive. An engineered asset is an investment because it is acquired with the expectation that it has the capacity to create value, or better still, that it will create value. The choice of acquiring the asset automatically imputes *ab initio* value to an asset. The choice, and hence the *ab initio* value, may be described in terms of opportunity cost (i.e., the benefit that could have been enjoyed if an alternative choice was made). The alternative value of giving up a benefit in order to acquire the productive capacity can be stated in terms of the quantitative cost incurred to acquire the asset. Indeed, the only reason for managing an asset is to realise value, and this, in the first instance, is the purpose and objective of the investment!

Paradoxically, an engineered asset requires upkeep, and poor upkeep can adversely affect the purpose of the investment, i.e., have an adverse impact on the creation of value. As illustrated in Fig. 2.2, the upkeep implies continuing investment, albeit that the magnitude and scale of the continuous injections of investment should be lower than the initial investment that created the asset in the first instance. Paradoxically, the upkeep investment injections are interpreted in terms of the cost doctrine. Thus, it is not surprising that emphasis is placed on the upkeep process which often manifests in terms of the dogma of cost reduction. The practical preference for the cost doctrine is probably because cost can be objectively estimated upfront via pricing and more accurately established *ex post facto*. However, investment appraisal is futuristic and involves a process of valuation, eventhough value is often established *prospectively*.

---

[5]Konstantin P., & Konstantin, M. (2018). Investment appraisal methods. In *Power and energy systems engineering economics*. Cham: Springer. https://doi.org/10.1007/978-3-319-72383-9_4.
  Götze, U., Northcott, D., & Schuster, P. (2008). *Investment appraisal: Methods and models.* Springer.

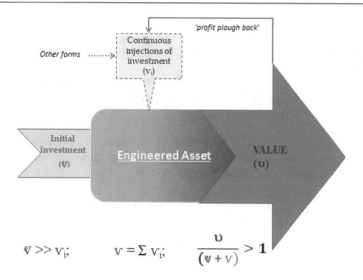

$$\mathfrak{v} \gg \mathrm{v_i}; \qquad \mathrm{v} = \Sigma\ \mathrm{v_i}; \qquad \frac{\upsilon}{\left(\mathfrak{v} + \mathrm{v}\right)} > 1$$

Fig. 2.2 An asset is an investment that generates value

2.3   Valuation

Valuation involves "an estimation of the worth of something, especially one carried out by a professional valuer". It is almost impossible to determine the value of an asset objectively because subjectivity is inextricable, even though the valuation of an asset is commonly stated in quantitative terms of the present value, that is, a date/time-stamped monetary equivalent representation of the asset's value. The subjectivity leads to many assumptions and scenarios, thus, in the traditional discipline of economics literature, valuation is regarded as an art, and the professional valuer has to comply with a variety of legal rules and standards.

The valuation of an asset depends (i) on the nature of the asset, (ii) how the asset is applied or deployed or what it is utilized for, (iii) the purpose or purposes of the valuation by the respective stakeholder(s) to the asset, and (iv) the particular point in time at which the valuation is determined.[6] There are many purposes for the valuation of an asset, e.g., for purposes of acquisition and deployment, auditing and financial accounting, insolvency and liquidation, insurance and warranties, invest- ment loans, taxation, or trade. The purpose of valuation and the life stage of an asset both determine the method that can be used to establish a valuation for the asset.

---

[6]National Association of Certified Valuation Analysts (www.nacva.com). Business valuation: fundamentals, techniques and theory.

Shapiro, E., Mackmin, D., & Sams, G. (1989). *Modern methods of valuation* (11th ed.). Routledge Taylor & Francis.

Interestingly, the aforementioned investment appraisal methods are also generally applied for valuation. There are three groups of internationally accepted approaches or methods for the valuation of fixed assets,[7] viz:

(i) The income approach. This is based on the earning capacity or income potential of the asset, that is, where the asset is utilized to generate income either directly or indirectly. This approach includes

   a. the more rigid capitalised earnings or capitalisation of cash flow method,
   b. the more flexible discounted cash flow method, and
   c. the economic value added (EVA) method.

For brevity, Eq. 2.1 provides a simplified expression for the (a) and (b) aspects of the income approach:

$$C_v = \frac{C_f}{P} * 100 \tag{2.1}$$

where C_v is the expected discounted return, C_f is the expected future return on assets (ROA), and P is the chosen interest rate in %.

More often, the income approach tends to be applied towards investment decisions such as the acquisition of an asset. In this regard, the cost of capital is a crucial factor and poor estimation of the cost of capital can lead to very unreliable valuation.[8] Furthermore, the reliability of the forecast cash flows is very dependent upon long term assumptions, future scenarios, and perceptions of future performance.

(ii) The second group of valuation methods refer to the market approach based on market comparisons or relative values. The market approach is generally more suited to determine the overall value of a business, especially to the valuation of intellectual property and intangible assets that can be sold or licensed. However, for an engineered asset, the valuation can be determined by comparative analyses of recent sales or purchases of the same or similar asset, assuming the existence of sufficiently reliable data. That is, the market approach involves an estimation of how a similar or a comparable asset is priced by the market. This approach not only subsumes the principle of substitution but also, the dogma is that the monetary worth of an engineered asset is what the market is willing to pay for the asset.

---

[7]Majtanova, A., & Sojkova, L. (2014). *Selected modern approaches to the valuation of fixed assets for insurance purposes.*

Tsamis, A., & Liapis, K. (2014). Fair value and cost accounting, depreciation methods, recognition and measurement for fixed assets. *International Journal in Economics and Business Administration, II*(3), 115—133.

[8]Sigurbjörn Haftórsson. (2015). Valuation Approaches in Practice. *Magister Scientiarum Thesis.* University of Akureyri.

Mathematically, the market approach can be expressed as

$$V = \frac{1}{n}\sum_{1}^{n} P_i * k_i \qquad (2.2)$$

where V is the market value of the asset, P_i is the purchase/selling prize, k_i is the coefficient of comparison, and n is the number of cases compared. In essence, the market approach determines a trade value of the asset at the particular point in time.

(iii)   A third grouping of valuation methods refers to the asset approach based on

   a. replacement cost of a new asset of the similar design, capability and capacity;
   b. 'fair value' based on exchange (i.e., similar to the market approach);
   c. 'good as new' valuation representing the original cost less depreciation plus maintenance costs;
   d. depreciated book value; and
   e. 'sum of parts' valuation.

For managers of deployed equipment and machinery, replacement[9] decisions due to degradation in performance tend to be the foremost purpose for valuation. The degradation in performance may manifest as technical malfunction or total functional failure. However, for structural assets (as predominant in infrastructure and facilities), refurbishment and renovation are also motivated by deterioration and/or obsolescence as well as concerns for the environment and ecology. The investment appraisal methods and the approaches to valuation lead to objective and quantitative descriptions of the value ethos in terms of efficiency as follows.

2.4 Value and Valuation

The relationship between value and valuation gives an indication of the rate at which an asset creates or destroys value. The ratio can be interpreted as efficiency, viz:

$$Asset\ efficiency = \frac{value\ obtained}{value\ invested} \qquad (2.3)$$

$$Value\ add\ efficiency = \frac{value\ obtained\ less\ value\ invested}{valuation\ of\ asset\ base} \qquad (2.4)$$

---

[9]Hartmann, J. C. *Engineering economy and the decision-making process.* Pearson. ISBN 9780131424012.

Calculating the efficiencies may not be trivial given that the value obtainable and value invested are cumulative over time and contain both quantitative and qualitative elements, albeit that the valuation of an asset can be established at a particular point in time.

The value obtained from an asset depends on the services demanded and the condition of the asset. The efficiency ratios have implications, especially with respect to the funding and/or financing for the acquisition/deployment, utilization and retirement of an asset. In essence, the value ethos can be resolved into (i) the value invested in an asset, (ii) the value obtained or obtainable from an asset, and (iii) the valuation of an asset. One reason for resolving the value ethos in this way is the desire to sustain the efficiencies in Eqs. 2.3 and 2.4 for as long as possible, at least, throughout the life of an asset.

2.5   Sustainability

According to Mahatma Ghandi, sustainability is founded on the existential realism "that there is enough for everyone's needs but not enough for everyone's greed." Therefore, in philosophical terms, 'sustainability' is an imperative of human existentialism. As depicted in Fig. 2.3, sustainability is about maintaining a dynamic equilibrium between people, prosperity, and the planet.

Sustainability has been defined in a number of domains,[10] for example, economic, material, life, social, or spiritual domain. For the management of an engineered asset, it is plausible to define sustainability by combining the following phrases:

  i. the ability to be maintained at a certain rate or level; with
 ii. the quality of being able to continue over a period of time;
iii. having minimum adverse impact on business, society and the environment; and
iv. maintaining ecological balance by avoiding the depletion of natural resources.

Hence, as a principle for managing assets, sustainability may be defined as:

the realisation of optimum value from engineered assets concurrently with resiliency to change stressors arising from the interactions between business, society, the environment and ecology. This implies the attainment of dynamic equilibrium between value, resiliency and adversity; in other words, a triangular equilibrium in the inextricable interactions between human endeavour/industry, human prosperity/well-being, and the planetary exigencies.

---

[10]Ben-Eli, M. (2015). *Sustainability: Definition and five core principles.* http://www.sustainabilitylabs.org/assets/img/SL5CorePrinciples.pdf.

Fig. 2.3 Sustainability
paradigm

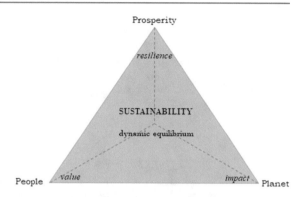

Sustainable asset management requires

i. that all activities and processes necessary to realise value from the asset are
   performed in dynamic equilibrium;
ii. that the profile of values specified by the various stakeholders of the asset can
    be dynamically optimised;
iii. concurrent minimisation of asset pollution of the environment;
iv. concurrent minimisation of adverse impact of the asset on ecology; as well as
v. ensuring that the asset remains resilient to endogenous and exogenous change
   stressors.

Items (iii) and (iv) describe the ecological footprint[11] of the asset.

An engineered asset such as a mobile phone is made from natural materials.
Despite the innate intelligence of human beings, however, at the end of the useful life
of the mobile phone, the material embodied in the phone may not be readily
decomposed to the natural state of the composite energy and matter in time to extract
and conflate similar amounts of energy and matter to manufacture another phone. The
manufacture of a new mobile phone often requires the extraction of more natural
resources, thus the grand challenge is to avoid further depletion of natural resources.
Requirements (i) and (ii) are related to the socio-economic domain whereas requirements (iii) and (iv) are more related to the material and 'life', i.e., ecological domain.

2.5.1 Socio-Economic and Ecological Sustainability

A depiction of sustainability in terms of socio-economic and ecological domains is
illustrated in Fig. 2.4. The ecological domain[12] describes nature's carrying capacity

---

[11]Where ecological footprint is measured in terms of (i) the quantity of nature used to produce the
asset, combined with (ii) the atmosphere and area of biologically productive land and water
required to assimilate the wastes generated by the asset.
[12]Braat, L. C., & Van Lierop, W. F. J. (1986). Economic-ecological modelling: an introduction to
methods and applications. *Ecological Modelling, 31*, 33–44.

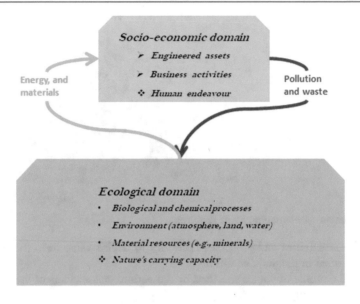

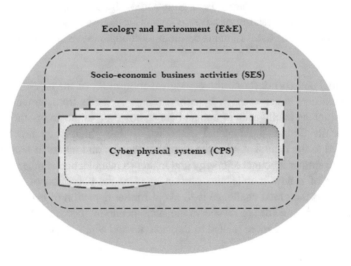

Fig. 2.4 Socio-economic and ecological domains of sustainability

in terms of atmospheric (i.e., air), land, and water environments, the associated biological and chemical processes, and corresponding material resources. On the one hand, energy and material resources extracted from the ecological domain are converted into engineered assets which support human existence and industry in the socio-economic domain. On the other hand, as the engineered assets are utilised to serve human existence and industry in the socio-economic domain, they cause pollution and waste to be generated into the ecological domain. An underlying premise of the Brundtland[13] definition for sustainable development is that the capacity of the ecological domain must remain much greater than that of the socio-economic domain so that the ecological processes perpetually sustains human existence and industry.

2.5.2 Sustainability and Asset Management

The management of engineered assets occurs within the socio-economic domain as depicted in Fig. 2.5. In this domain, businesses deploy engineered assets to provide goods (products) and services that serve human livelihood, support industry and human existence. In return, consumers of the goods and services pay for and motivate business enterprise. In as much as the engineered asset provides the means for a business organisation to realise value, it is also the engineered asset that not only links, but also, provides the means to sustain the relationship between business and customer. Thus, the circular loop depicted in the socio-economic domain implies that management of engineered assets is subjected to the sustainability imperative and conforms to the circular economy[14] model.

Ideally, the sustainability imperative demands that engineered assets that have already been transformed from the ecological domain perpetually remain in the socio-economic domain. In practice, the need to avoid scarcity by minimising the rate of depletion of natural resources requires that engineered assets are sustained in the socio-economic domain for the longest possible time. Furthermore, engineered assets must be managed so as to minimise the adverse effects of the consequential pollution and waste on the environment. Of course, the optimism is that the ecological domain can safely absorb the pollution and waste such that there are minimal interferences, interruptions or disruptions of the natural cycles of the biological and chemical processes. The sustainability imperative applies irrespective of the industry sector. The following examples buttress the point.

A floating production, storage and offloading (FPSO) vessel used in the offshore oil and gas sector typically comprises infrastructure, facilities, equipment, machinery, subsystems and systems for the extraction, processing, storage, and transfer (offloading) of hydrocarbons. The operator of the FPSO deploys the wide ranging composition of assets to serve its customers. Similarly, a rail operator deploys track and rolling stock assets to serve its customers' transportation needs; the same applies to an electricity utility, a manufacturer of fast-moving consumer goods, a miner, a water utility, a provider of information, computing and

---

[13]Brundtland Report. (1987). 1983 World Commission on Environment and Development. www.sustainabledevelopment2015.org/.../our-common-future.

[14]A circular economy is an industrial system that is restorative or regenerative by intention and design. It replaces the end-of-life concept with restoration, shifts towards the use of renewable energy, eliminates the use of toxic chemicals and aims for the elimination of waste through the superior design of materials, products, systems and business models. Nothing that is made in a circular economy becomes waste, moving away from our current linear 'take-make-dispose' economy.

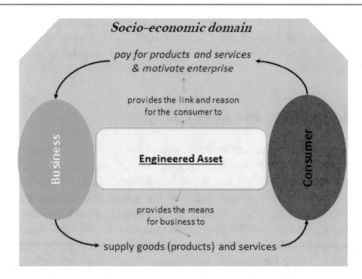

Fig. 2.5 Engineered assets within circular model of socio-economic sustainability

telecommunications services, et cetera. In each case, the engineered assets link, connect and underpin the consequent relationship between the business organisation and the client/customer. Thus, engineered assets are central to the sustainability of the socio-economic domain and the circular economy model.

The sustainability principle demands sustained capturing of value from existing engineered assets concurrently with minimising or even eliminating, as well as being resilient to the impacts of pollution, waste, supervening forces, and climate change effects. In the simplest terms, the ideal is to 'design out' toxins and congestion.[15] In practice, the requirements for sustainable asset management can be effected through techniques that incorporate dematerialisation, life-extension, reduction, reuse, and recycling concepts. The techniques focus on retaining the asset within the socio-economic domain for the maximum possible life time, so the techniques depend on definitions of asset life.

(a) *Dematerialisation (see also Chap. 8)*

Dematerialisation focuses on avoiding the depletion of natural resources by:

(i) converting functions performed using corporeal (tangible) assets into functions that can be performed via incorporeal (intangible) assets;
(ii) creating engineered assets that offer multi-functional capabilities.

---

[15]Kemp, L. (2019) The circular economy has become the norm: A vision for 2030. World Economic Forum 29 Oct 2019.

In a sense, dematerialisation has been a driver for technological innovations that underpin the multi-functionality common in modern corporeal assets. Where feasible, instead of creating an engineered asset from natural resources when a new function is required, an existing asset may be 'configured' or 'programmed' or even re-designed to perform additional functions. In fact, mobile phones are being utilised to perform a number of functions, re: communication, data storage, information exchange, photography (camera), et cetera.

(b) *Reduction*

Reduction techniques focus on containment of the asset's ecological footprint by minimising the impact of the asset on the environment and ecology, e.g., reductions in the material content of an asset, the energy consumption of an asset, as well as pollution and waste emanating from an asset. Hence, the determination of the energy efficiency of the asset is a crucial management activity. This requires accurate and dynamic measurements of the relevant parameters.

(c) *Recycling*

Recycling is based on the premise that, when an asset has reached the end of its current functional life, it is possible to decompose, then, convert the composite energy and materials to create a new but different asset. Again, recycling is aimed at containment of the asset's ecological footprint. This means that the material and energy composition of the asset are not removed from the socio-economic domain by being converted into pollution and waste. An ultimate goal of recycling should be to perpetually retain the material composition of an asset within the socio-economic domain. A valuation of the asset may be carried out for the purpose of recycling.

(d) *Re-use*

Re-use is premised on the assumption that the function that an asset performs is perpetual, for example, the cleaning of a ballast to make it ready for reuse, or erasing of a memory device so that old and/or new data can be stored and retrieved, or reconfiguring a passenger aircraft for freight purposes. Thus, when the functional use of an asset expires as deployed in one context, the same asset can be redeployed in another context to perform its original or similar function. The challenge is to ensure that in each context that the asset is re-used, the functional application of the asset also minimises environmental impact and ecological footprint. Trade and insurance valuations may be carried out for the purpose of re-using the asset in the same functional mode.

(e) *Life extension*

Perhaps, the most pragmatic approach is to extend both the corporeal and inherent functional life of an asset. If an asset is a motorised vehicle, then repairing the engine by replacing a failed component, or upgrading the braking system, or replacing a worn tyre may be regarded as tasks that extend the life of the vehicle. Thus, it is not surprising that extant literature[16] abounds on techniques for extending the useful life of an engineered asset. Life extension depends on:

   i. how an asset is used and the conditions within which the asset is utilised,
  ii. the monitoring of condition and tracking of reliability so as to implement correct maintenance interventions, plus,
 iii. holistic assessment of the overall asset condition so that investments are appropriately prioritised to assure the desired service delivery capability of the asset.

2.6 Summary of the Chapter

This chapter introduced the value ethos, together with investment and valuation concepts, and sustainability as fundamental principles for managing engineered assets. Methods for investment appraisals were highlighted in conjunction with approaches to valuation. The significance of the circular economy model was mentioned in the context of sustainability. The operationalization of sustainability, that is, sustainable asset management techniques such as de-materialisation, recycling, reduction, reuse, and life extension were discussed.

---

[16]Woodhouse, J. (2012). Making the business case for asset life extension. In *IET and IAM Asset Management Conference 2012*. https://doi.org/10.1049/cp.2012.1900.

Norris, M. (2013). Planning to extend the life of major assets. *Procedia CIRP, 11*, 207–212.

Khatib, Z., & Walsh, J. M. (2014). Extending the life of mature assets: How integrating subsurface and surface knowledge and best practices can increase production and maintain integrity. SPE-170804-MS. https://doi.org/10.2118/170804-MS.

Madusanka, W. M. L., Rajini, P. A. D., & Konara, K. M. G. K. (2016). Decision making in physical asset repair/replacement: A literature review. In *13th International Conference on Business Management (ICBM)*, University of Sri Jayewardenepura, Sri Lanka, Available at SSRN https://ssrn.com/abstract=2910207 or https://doi.org/10.2139/ssrn.2910207.

International Transport Forum. (2018). Policies to extend the life of road assets. https://www.itf-oecd.org/sites/default/files/docs/policies-extend-life-road-assets.pdf.

Gavrikova, E., Volkova, I., & Burda, Y. (2020). Strategic aspects of asset management: an overview of current research. *Sustainability 12*, 5955. https://doi.org/10.3390/su12155955.

2.7 Exercises

1. With regard to the management of assets, why is value described as an ethos in this book?
2. The Covid-19 pandemic is transforming livelihoods, industry and society. Use one or two examples to discuss the influence of the pandemic on sustainable management of engineered assets.

References and Additional Reading

Ben-Eli, M. (2015). *Sustainability: Definition and five core principles.* http://www.sustainabilitylabs.org/assets/.

Braat, L. C., & Van Lierop, W. F. J. (1986). Economic-ecological modelling: An introduction to methods and applications. *Ecological Modelling, 31*(1986), 33–44.

Brundtland Report. (1987). 1983 World Commission on Environment and Development. www.sustainabledevelopment2015.org/…/our-common-future.

Gavrikova, E., Volkova, I., & Burda, Y. (2020). Strategic aspects of asset management: An overview of current research. *Sustainability, 12*, 5955. https://doi.org/10.3390/su12155955.

Götze, U., Northcott, D., & Schuster, P. (2008). *Investment appraisal: Methods and models.* Springer.

Haftórsson, S. (2015). Valuation approaches in practice. *Magister Scientiarum Thesis.* University of Akureyri.

Hartmann, J. C. (2006). *Engineering economy and the decision-making process.* Pearson. ISBN 9780131424012.

International Transport Forum. (2018). Policies to extend the life of road assets. https://www.itf-oecd.org/sites/default/files/docs/policies-extend-life-road-assets.pdf.

Kemp, L. (2019). *The circular economy has become the norm: A vision for 2030.* World Economic Forum 29 Oct 2019.

Khatib, Z., & Walsh, J. M. (2014). Extending the life of mature assets: How integrating subsurface and surface knowledge and best practices can increase production and maintain integrity. SPE-170804-MS. https://doi.org/10.2118/170804-MS.

Kilmann, R. H. (1981). Toward a unique/useful concept of values for interpersonal behaviour: A critical review of the literature on value. *Psychological Reports, 1981*(48), 939–959.

Konstantin, P., & Konstantin, M. (2018). Investment appraisal methods. In *Power and Energy Systems Engineering Economics.* Springer, Cham. https://doi.org/10.1007/978-3-319-72383-9_4.

Madusanka, W. M. L., Rajini, P.A. D., & Konara, K. M. G. K. (2016). Decision making in physical asset repair/replacement: A literature review. In *13th International Conference on Business Management (ICBM)*, University of Sri Jayewardenepura, Sri Lanka. Available at SSRN https://ssrn.com/abstract=2910207 or https://doi.org/10.2139/ssrn.2910207.

Majtanova, A., & Sojkova, L. (2014). *Selected modern approaches to the valuation of fixed assets for insurance purposes.*

National Association of Certified Valuators and Analysts. (1995–2012). *Business valuation: Fundamentals, techniques and theory.* http://edu.nacva.com/preread/2012BVTC.

Norris, M. (2013). Planning to extend the life of major assets. *Procedia CIRP, 11*(2013), 207–212.

Pauls, R. (1990). Concepts of value: A multidisciplinary clarification. ISBN 1-86931-036-5.

Shapiro, E., Mackmin, D., & Sams, G. (1989). *Modern methods of valuation* (11th ed.). Routledge Taylor & Francis.

Sidorchuk, R. (2015). The concept of value in the theory of marketing. *Asian Social Science, 11*(9). ISSN 1911-2025.

Tsamis, A., & Liapis, K. (2014). Fair value and cost accounting, depreciation methods, recognition and measurement for fixed assets. *International Journal in Economics and Business Administration, II*(3), 115–133.

Woodhouse, J. (2012). Making the business case for asset life extension. In *IET and IAM Asset Management Conference 2012*. https://doi.org/10.1049/cp.2012.1900.

Technical Principles

3

Abstract

Many scholars and practitioners whose backgrounds are in the engineering and associated technical disciplines tend to focus on the utilisation stage and thus approach the management of engineered assets from the stance of risk, reliability, and cost. Given this stance, operators and maintainers commonly focus on safety risks as well as risk of malfunction or functional failure of a component, spare part, subassembly, subsystem or the whole asset. The era of Society 5.0 features increasing dependence on hyper-interconnected and interdependent cyber physical systems of engineered assets that are confronted by stressors attributable to VUCA phenomena like the Covid-19 pandemic, climate change, or *vis major* events, that is, → superior powers or forces which can neither be resisted nor controlled. Therefore, the concepts of resilience and vulnerability have become significantly pertinent for managing interconnected systems of engineered assets, especially because CPSs are embedded within socio-economic systems (SES), and both CPS and SES are encapsulated within the environment and ecology. The discourse here highlights the concepts of reliability, risk, resilience, vulnerability, and condition from a pragmatic viewpoint. Readers interested in detailed treatise of these concepts may refer to other extant literature. Philosophically, these concepts are rooted in the quest of how to deal with uncertainty.

3.1 Uncertainty

According to the dictionary, uncertainty represents:

- a situation in which something is not known, or
- where the order or nature of things is not known, or

- where the consequences, extent, or magnitude of circumstances, conditions, or events are unpredictable, and mode of occurrence of something is not known;
- a situation in which credible probabilities to possible outcomes cannot be easily assigned;
- the degree to which available choices or the outcomes of possible alternatives are free from constraints.

Absolute uncertainty has no historical antecedent. This means that there is lack of data, information, and knowledge, and/or inadequate skill and competency to deal with an absolutely uncertain condition, circumstance, event, phenomenon, situation, or something. It is pragmatic to perceive uncertainty as arising from the dicephalous nature of *vis major* (uncontrollable and unavoidable superior forces) and *casus fortuitus*[1] (unforeseeable and unforeseen) VUCA events. The unavoidable, uncontrollable and unforeseeable nature of uncertainty means that the effects of the corresponding stressors are non-maskable.

Figure 3.1 represents a mind mapping of uncertainty, risk and opportunity as a spectrum. Ontologically, uncertainty can be viewed either from a positivistic or negativistic mind set or perspective. On the one hand, a positivistic perception of uncertainty can engender optimism that encourages the exploration of valuable opportunities (i.e., moving towards the realm of 'unknown knowns'). On the other hand, a negative perception of uncertainty can conjure pessimism, desperation or hopelessness towards the realm of 'unknown unknowns'. The negativistic mind set may also lead to uncertainty being perceived as the source of threat(s)[2] to something of value. It is remarkable that in management parlance, the spectrum ranging from uncertainty, through risk to opportunity is conventionally analysed in terms of strengths, weaknesses, opportunities, and threats (SWOT).

With regard to engineered assets, the different and vagarious value propositions of the stakeholders represent sources uncertainties, risks and opportunities that influence management decision making. The management challenge is to synthesise the uncertainty, risk and opportunity spectrum to facilitate rational decision making.[3] Thus, for technical managers of engineered assets, it is not surprising that the concept of risk is often the starting point.

---

[1] *casus fortuitous* → an exceptional or extraordinary occurrence that is not reasonably foreseeable.
[2] Johansen, A., Halvorsen, S. B., Haddadic, A., & Langlo, J. A. (2014). Uncertainty management—A methodological framework beyond "The Six W's". *Procedia—Social and Behavioral Sciences, 119*, 566–575.
[3] Beedles, M. (2017). *Asset management for directors*. Australian Institute of Company Directors. ISBN 9781876604394.

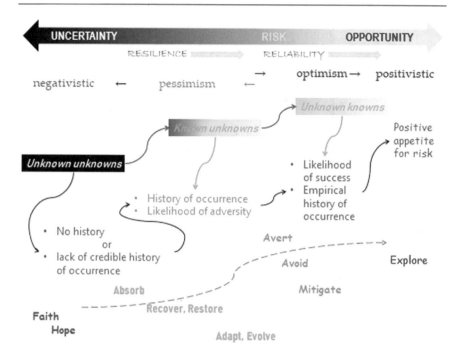

Fig. 3.1 Uncertainty, risk and opportunity

3.2 Risk

A ramification of the pessimistic mind set is that natural events like earthquakes, floods, and lightning are typically considered to be sources of negative uncertainty because the corresponding stressors often cause adverse effects or result in undesirable or 'abnormal' consequences. Furthermore, human behaviour such as violent disturbances, riots, terrorism, and the worst case scenario of wars also cause damage and destruction of engineered assets. All over the world, there are daily accounts of disruption, damage, and destruction of engineered assets caused by such natural, indirect and direct human-induced events.[4] With regard to managing engineered assets, the pragmatic approach considers such uncertain events as exogenous sources of risk, albeit that an asset naturally contains endogenous sources of risk. This endogenous aspect will be discussed further as part of the concept of vulnerability.

Interestingly, ISO Guide 73 defines risk 'as the probability that exposure to a hazard will lead to a negative consequence; alternatively, a hazard poses no risk if there is no exposure to that hazard.' ISO 31000[5] further defines risk as the negative

---

[4]www.ranker.com/list/worst-natural-disasters-2019/ranker-news.
[5]ISO 31000:2009. Risk management—Principles and guidelines.

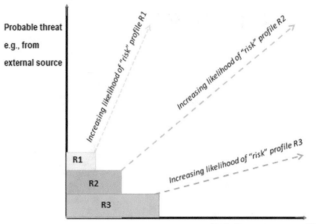

Fig. 3.2 Classical two-dimensional model of risk

'effect of uncertainty on objectives'; this implies the interpretation or transformation of uncertainty into threats. Although stressors arising from uncertain events can
create opportunities, however, the negativistic convention is that such stressors
cause deviations from expectations and threaten objectives. Thus, the
two-dimensional models of risk depicted in Fig. 3.2 focuses on the happenstance of
threats, and the likely consequences of the happenstance of threats.

By this convention, risk management involves identification, evaluation and
prioritization of threats (i.e., sources of risks) followed by coordinated application
of knowledge, skills, competencies, and where necessary, material resources to
monitor the probabilities, to avert, avoid, and/or to mitigate and control the consequences of the threats. Considering that a threat to an asset represents a source of
risk, it follows that managing risk[6] is an implicit task of asset management. After
all, any source of risk represents a threat to something of value.

The negativistic ontology of uncertainty also leads to the consideration of risk in
terms of 'criticality', i.e., in terms of the relationship between the frequency of
failure/fault occurrence and the severity of the failure/fault. This pragmatic
approach (see Fig. 3.3) provides a means for the ranking of risks, facilitates the
prioritization of risks, and informs decision-making based on the ranking and prioritization of risks.

[6]Davis, R. (2017). An introduction to asset management. ISBN 978-0-9571508-3-6.

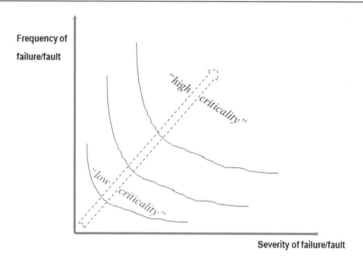

Fig. 3.3 Classical two-dimensional model representation of criticality

3.3   Reliability

As a derivative of risk, the concept of reliability is extensively covered in literature and many standards,[7] however, for brevity, we shall define reliability as

> the chance (probability, likelihood) that an engineered asset will function as intended, under reasonably specified conditions, for a specified time period, and without suffering any pre-defined malfunction or failure.

This means that to effectively manage the reliability of an asset, there are five lemmas to consider. The first concerns the *functioning* of an asset, that is, how an asset is deployed and utilized. Deployment and utilisation of an asset are automatically derived from the value to be appropriated from the asset, that is, the asset has to be deployed and utilised for a defined purpose. Ideally, the value or purpose should be defined from the conflated aspirations of the stakeholders. The common quandary or practical dilemma is that the purpose of an asset is typically defined with preference to the interests of the investor/shareholder rather than the conflated interests of all the stakeholders.

The second lemma is that the failure characterisation of the asset must be pre-defined and agreed *ab initio*. Such characterisation may range from technical malfunctioning (soft failure) to total functional or hard failure. A common reason for poor decision making in asset management is ambiguity, vagueness or lack of clarity but especially, lack of *apriori* agreement on the failure characterisation of an asset. For instance, the maintainer of an asset may be more concerned by the

---

[7]Menčík, J. (2016). Standards related to reliability. https://doi.org/10.5772/62366.

slightest suggestion of the likelihood of failure as indicated from the signals being monitored, whereas the operator is more likely to be convinced by the functional or hard failure of the asset.

The third lemma is pre-specification of the time interval or period during which the reliability of an asset is considered. A common misnomer is to count the number of times that a component, spare part, subsystem, or an asset has failed without pre-specification of the time interval or period. This misnomer not only gives rise to inaccurate measurements and reckless recording of time between failures but also, just counting the instances of malfunction or failure often misleads decision making. Furthermore, inaccurate measurements or reckless recording of time intervals vitiates mathematical[8] calculations of reliability; for an asset that has failed three times within a stated period, it does not make sense to speak about mean time between failures (MTBF) since the data is non-statistical!.

The fourth lemma is pre-specification of the range of conditions or the operating regime(s) within which an asset should function or perform as intended. It is ambiguous to indicate that an asset has malfunctioned or failed without a succinct description of the corresponding conditions within which the malfunction or failure occurred. For example, assume that skidding is a failure characterisation of a motorised vehicle (e.g., a motorcar). This malfunction is more likely to occur whilst the vehicle is driven on a wet road under rainy conditions, and may subsequently result in a hard failure of the vehicle. When the vehicle is driven under rainy conditions, the relative criticalities of the components are not the same as when the car is driven under non-rainy conditions. A rainy and wet road condition is an exogenous source of risks (e.g., skidding, poor visibility, et cetera) that affect the operational reliability of the vehicle. The probability that the driven vehicle will skid is more or less an estimation of uncertainties associated with rainy and many other conditions that include different types of road pavements and other road users. Incidentally, the failure modes inherent in a motorised vehicle are also endogenous sources of uncertainty. The preceding four lemmas highlight the distinction between inherent and operational reliability, and the combination into a fifth lemma leads to the concept of asset integrity as follows.

3.4 Asset Integrity

An important concept that derives from the synthesis of risk, criticality, and reliability is asset integrity. Whereas the common terminology is technical integrity,[9] however, asset integrity may be defined as

---

[8]IEC 61703:2016 Standard | Mathematical expressions for reliability, availability, maintainability and maintenance support terms.
[9]Technical integrity—attributes that reflect the intended, desired emergent properties of a system and the minimisation of unintended, undesired emergent properties.

the chance that an asset will continue to provide the means for the realization of value without destroying the intrinsic or inherent value of the asset, and without imposing unforeseeable adverse effect on business, people, society, ecology and the environment.

From a management point of view, asset integrity demands that the concepts of risk and the sustainability principle are fused within the value ethos. As a corollary, asset integrity reiterates that safety is embedded in both the sustainability paradigm and the value ethos, irrespective of whether or not an asset is vulnerable to internal or external sources of threats.

3.5 Vulnerability

Vulnerability extends the concept of risk by focusing on the exposure[10] and susceptibility of an asset to a threat. Thus, vulnerability is about the weaknesses inherent in an asset. An asset (or anything, really) is regarded as being vulnerable if its inherent or intrinsic weaknesses are not protected, which means that an asset is exposed to the possibility of being 'attacked'. With reference to the sustainability paradigm,[11] vulnerability may be defined as the degree to which an asset is likely to malfunction or fail when exposed to a particular hazard, and where the hazard imposes a particular form of perturbation and stressor.

Vulnerability is related to reliability in the sense that the least reliable component (s) of an engineered asset can make the asset susceptible to particular threats or forms of perturbations and stressors. The susceptibility attributable to the least reliable part(s) also diminishes the integrity of an asset.

Somewhat as a corollary to risk management, vulnerability assessment involves identifying weaknesses inherent in an asset, monitoring the susceptibility of the weaknesses to various internal and external threats, and providing necessary knowledge for coordinated application of resources, either to prevent an attack(s), or to protect the asset from exposure to particular forms of attack, or to mitigate the consequences of unfortunate attack(s) on the weaknesses inherent in the asset. The pragmatic way to deal with vulnerability is to protect an asset presuming that particular threats can materialise, i.e., there is reasonable likelihood of particular hazards, i.e., forms of perturbations and stressors.

The concept of vulnerability more or less derives from the negativistic perception of uncertainty. In practice, the management of vulnerability focuses on protecting an asset against *casus fortuitous* threats often associated with wanton human behaviour such as violent disturbances, riots, terrorism and, where possible, from the effects of a war. In this sense, vulnerability is a twin concept of resilience, particularly because the conventional focus of resilience is on the stressors arising from *vis major* events or supervening uncertainty.

---

[10]Boyce, R. (2019). *Vulnerability assessments: The pro-active steps to secure your organisation.* SANS Institute.

[11]Moret, W. (2014). Vulnerability assessment methodologies: A review of the literature. A Report for USAID ASPIRES PEPFAR fhi360.

3.6 Resilience

Philosophically, the concept of resilience (or resiliency) is based on the consideration of the supervening possibility of uncontrollable and unforeseeable superior forces. As depicted in the mind map of Fig. 3.1, the concept engenders a positive approach to dealing with uncertainty. Increasing instances of VUCA phenomena have also renewed emphasis on the concepts of resilience and vulnerability. It is arguable that there is the paramount need to focus on resilience given our irrevocable dependence on the cyber physical conflation of systems of engineered assets in all aspects of human endeavour, livelihood, and industry.

Cyber physical systems are not only composed of highly automated engineered assets but also, the systems are highly interdependent because of the hyper connectivity that exists between the composite engineered assets. Data and the internet enable the coordination, control, monitoring and operation of complex gadgets, equipment, machinery and the interdependent infrastructure systems. The need to 'manage' resilience (and vulnerability) arises from the fact that the malfunction or failure of a component or a subsystem of an asset may readily cascade into malfunctions and failures in other components and subsystems of other interconnected assets. Cascaded malfunctions or failures of components and subsystems of interconnected assets can, worse still, (i) trigger catastrophic failure of whole CPS,[12] (ii) cause cataclysmic disruption to socio-economic activities, as well as (iii) stimulate events that can result in devastation on the natural environment.

The concept of resilience is conventionally considered on the basis that *vis major* natural phenomena (like earthquakes, floods, and lightning) tend to disrupt and degrade the performance of engineered assets, and worse still, damage engineered assets. The adverse effects of *vis major* or *casus fortuitus* change stressors on engineered assets are often exacerbated by poor design, construction, installation, commissioning, and abusive utilisation of an asset. For this reason, resilience engineering focuses on creating inherent robustness within an asset.

Broadly speaking, resilience[13] refers to:

- the capacity to recover quickly from adverse effects;
- "the ability to prepare for threats, coupled to the ability of a system to absorb impacts, recover and adapt following persistent stress or a disruptive events"[14];

---

[12]Alhelou, H. H., Hamedani-Golshan, M. E., Njenda, T. C., & Siano, P. (2019) A survey on power system blackout and cascading events: Research motivations and challenges. *Energies 12*, 682. https://doi.org/10.3390/en12040682.

[13]Southwick, S. M., Bonanno, G. A., Masten, A. S., Panter-Brick, C., & Yehuda, R. (2014). Resilience definitions, theory, and challenges: interdisciplinary perspectives. *European Journal of Psychotraumatology, 5*, 25338.

[14]Marchese, D., Reynolds, E., Bates, M. E., Morgan, H., Clark, S. S., & Linkov, I. (2018). Resilience and sustainability: Similarities and differences in environmental management applications. *Science of the Total Environment, 613–614* 1275–1283.

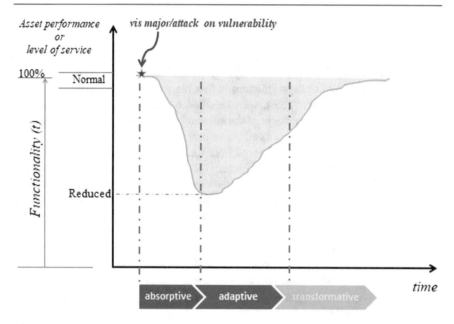

Fig. 3.4 Three stages of resilience

- "the ability to recover quickly and resume original level of service after damage"[15];
- "the ability to prepare for, and adapt to changing conditions, plus withstand and recover rapidly from disruptions";
- "the ability to prepare and plan for, absorb, recover from, and more successfully adapt to adverse events".

In this book, we adopt a much broader definition of resilience as *the capability to sustainably accommodate the effects of change inducing stressors.*[16] This definition leads to the resolution of resiliency into three stages as illustrated in Fig. 3.4. The first stage assumes that the change inducing stressors may only be transient. Hence, following the incidence of transient change stressor(s), the asset (CPS) may enter into a state in which it absorbs the energy induced by the perturbation or transient disruptive force(s). This positivistic or optimistic view of uncertainty manifesting as transient stressors is that the absorptive resilience stage includes phases of resistance, recovery and possible restoration of critical elements of an asset, i.e., the

[15]Brabhaharan, P. Resilience planning in road network asset management. In *IPWEA Conference*, Wellington, 7–8 November 2016.
[16]Amadi-Echendu, J. E., & Thopli, G. A. (2020). Resilience is paramount for managing socio-technological systems during and post Covid-19. *IEEE Engineering Management Review*, 48(3).

'bounce back' phases, albeit that the likelihood of a 'bounce back' depends on the magnitude of the transient change stressors.

The second stage of resiliency derives from the fact that the change inducing stressors may be intransigent. Similarly, the positivistic view of uncertainty manifesting as intransigent change stressors is that the asset (CPS) transitions into an adaptive stage of resiliency. The adaptive resilience stage encompasses the capacity to absorb the transient components of the intransigent change stressors concurrently with adaptation to energies due to all prevailing stresses.

The third stage of resiliency derives from the fact that the nature of the change inducing stressors may also be evolutionary. If so, the positivistic view of uncertainty manifesting as evolutionary change inducing stressors is that the asset (CPS) should possess the capability to transform accordingly. That is, the asset (CPS) should be capable of continuous adaption to the changes induced by evolutionary stressors.

Resilience is related to reliability in the sense that the most reliable component(s) provide robustness (strength) in the asset's ability to withstand disruptions. From a management viewpoint, resilience can be characterised by the closely related terms of (i) redundancy, (ii) robustness, (iii) recoverability, (iv) rapidity, (v) resourcefulness, and (vi) adaptability. Robustness may not only be established in the form of hardware redundancy, but also, redundancy may be created in a 'soft' form. Nevertheless, it is relatively easier to articulate robustness or to establish redundancies during the creation of an asset, and this is often the purview of reliability and resilience engineering.

Preliminary assessment[17] of resilience involves identification of the strengths in terms of inherent redundancies or robustness of an asset. This can be achieved proactively during the design, manufacturing, construction, installation, and commissioning of an asset (CPS). Otherwise, during the utilisation stage in the life of an asset, detailed assessment of resilience involves retrospective identification of redundancies and weaknesses (vulnerabilities) associated with an asset plus prospective estimation of the likelihood of occurrence of *vis major* or *casus fortuitus* phenomena.

Furthermore, it is possible to develop theoretical models of resilience through empirical determination of the parameters highlighted in Fig. 3.5. The development of such a model not only provides necessary knowledge but also, may be used to simulate coordinated application of resources that can mitigate the consequences of an unfortunate *vis major* and/or *casus fortuitus* phenomena.

Whereas robustness and redundancy are often associated with reduced or partial loss of asset functionality, however, recoverability and rapidity deal with both partial and total loss of asset functionality. Resourcefulness is about the availability

---

[17]Liu, W. (2014). The application of resilience assessment—Resilience of what, to what, with what? A case study based on Caledon, Ontario, Canada. *Ecology and Society, 19*(4), 21. http://dx.doi.org/10.5751/ES-06843-190421.

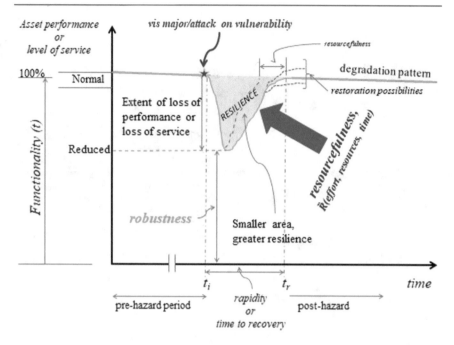

Fig. 3.5   Modelling resilience of an asset (CPS) in terms of functionality/performance

of complementary resources that enable recoverability and facilitate rapidity (restoration). In a sense, both resourcefulness and adaptability take into account the social dimension implicit in a CPS. Adaptability refers to the ability to adjust to new conditions. Thus, the resiliency of an asset encompasses the ability to recover from disruptive attacks to include adaptability to evolving conditions (uncertainties).

Figure 3.6 provides an illustration of how the technical principles of uncertainty, risk, reliability, integrity, resilience and vulnerability may be integrated to manage assets. Reliability plus risk modelling and analyses may focus at the component and subsystems levels of the asset hierarchy, whereas resilience and vulnerability modelling and analyses may be focused at the systems levels of the asset hierarchy. The picture in Fig. 3.6 shows that the inherent redundancies, vulnerabilities and responsiveness should be identified so as to establish the levels of robustness and resourcefulness necessary for sustainable management of an asset or the agglomeration of assets as in a CPS. Given the coupling of biological systems, the technical and analytical models must consider that socio-economic and ecological/environmental stressors inadvertently influence the functioning and performance of asset or the agglomeration of assets as in a CPS. Hence, resilience management incorporates systems dynamics modelling to account for the inextricable interrelationships between the behavioural elements of the socio-economic and socio-political environments within which the CPS is embedded.

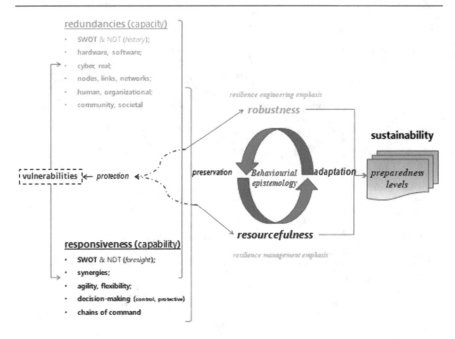

Fig. 3.6 A conceptual model for managing risk, resilience, and vulnerability of asset systems

3.7   Asset Condition

Technically, the *condition* of an asset reflects how the principles of uncertainty, risk, reliability, integrity, resilience and vulnerability are being integrated to manage the asset. From a pragmatic viewpoint, once an asset has been established and deployed for purpose, then, on-going management of the asset predominantly depends on the *condition* of the asset. In other words, the *condition* of the asset is the paramount determinant of how the asset is managed. The implication here is that the *condition* of an asset encompasses a number of dimensions and perspectives that facilitate robust decision making. After all, an asset that is not delivering value (or destroying value) cannot be considered to be in its best possible condition, notwithstanding how value is measured.

Often, based on respective stakeholder persuasions and discipline preferences, the value and consequently, the condition of an asset may be narrowly or widely stated. Invariably, a narrow statement of value (e.g., purely in economic/financial dimensions) may also lead to a vitiated statement of the condition of an asset. For instance, suppose that an asset provides good economic profit whilst concurrently producing waste and/or polluting the environment. How should the condition of the asset be stated? A narrow or poor assessment of the condition of an asset may subsequently result in short term approaches that exacerbate suboptimal decision making. In subsequent chapters, we will discuss how the condition of an asset should be assessed in at least five contexts, viz—ecological/environmental, economic/financial, functional, technical/technological and socio-political.

3.8 Summary

The discourse in this chapter shows that how to deal with uncertainty provides the primal motive for asset management. The pragmatic approach is to resolve uncertainty into risk, reliability, integrity, resilience and vulnerability as the technical principles for managing engineered assets. Whereas the principles of risk, reliability, and integrity may be readily applied to manage discrete or single assets, however, the principles of resilience and vulnerability extends to the management of systems of assets or cyber physical systems.

3.9 Exercises

Choose an asset that has been operational for at least 5 years.

1. Collect appropriate data and information to estimate the following

   i. reliability,
   ii. criticality, and
   iii. robustness of the asset.

2. Propose how the resilience and vulnerability of the chosen asset should be managed.

References and Additional Reading

ADB & GCA. (2021). A system-wide approach for infrastructure resilience. Technical Note: 2021 Asian Development Bank (ADB) and Global Center on Adaptation (GCA). ISBN 978-92-9262-694-5. http://dx.doi.org/10.22617/TCS210017-2

Alhelou, H. H., Hamedani-Golshan, M. E., Njenda, T. C., & Siano, P. (2019). A survey on power system blackout and cascading events: Research motivations and challenges. *Energies, 12,* 682. https://doi.org/10.3390/en12040682.

Amadi-Echendu, J. E., & Thopli, G. A. (2020). Resilience is paramount for managing socio-technological systems during and post Covid-19. *IEEE Engineering Management Review, 48*(3).

Beedles, M. (2017). *Asset management for directors.* Australian Institute of Company Directors. ISBN 9781876604394.

Béné, C., Godfrey-Wood, R., Newsham, A., & Davies, M. (2012). *Resilience: New Utopia or New Tyranny? Reflection about the potentials and limits of the concept of resilience in relation to vulnerability-reduction programmes* (Vol. 2012(405), pp. 1–61). IDS Working Paper.

Boyce, R. (2019). *Vulnerability assessments: The pro-active steps to secure your organisation.* SANS Institute.

Brabhaharan, P. (2016) Resilience planning in road network asset management. In *IPWEA Conference*, Wellington, 7–8 November 2016.

Cançado, D., & Mullan, M. (2020). *Stock-take of climate resilient infrastructure standards.* Working Paper: Global Center on Adaptation. www.gca.org

Davis, R. (2017). An introduction to asset management. ISBN 978-0-9571508-3-6.

Hall, J. W., Aerts, J. C. J. H., Ayyub, B. M., Hallegatte, S., Harvey, M., Hu, X., Koks, E. E., Lee, C., Liao, X., Mullan, M., Pant, R., Paszkowski, A., Rozenberg, J., Sheng, F., Stenek, V., Thacker, S., Väänänen, E., Vallejo, L., Veldkamp, T. I. E., van Vliet, M., Wada, Y., Ward, P., Watkins, G., & Zorn, C. (2019). *Adaptation of infrastructure systems: Background paper for the global commission on adaptation.* Oxford: Environmental Change Institute, University of Oxford.

IEC 61703:2016 Standard. Mathematical expressions for reliability, availability, maintainability and maintenance support terms.

ISO 31000:2009. Risk management—Principles and guidelines.

Johansen, A., Halvorsen, S. B., Haddadic, A., & Langlo, J. A. (2014). Uncertainty management—A methodological framework beyond "The Six W's". *Procedia—Social and Behavioral Sciences, 119*(2014), 566–575.

Liu, W. (2014). The application of resilience assessment—resilience of what, to what, with what? A case study based on Caledon, Ontario, Canada. *Ecology and Society, 19*(4), 21. https://doi.org/10.5751/ES-06843-190421.

Marchese, D., Reynolds, E., Bates, M. E., Morgan, H., Clark, S. S., & Linkov, I. (2018). Resilience and sustainability: similarities and differences in environmental management applications. *Science of the Total Environment, 613–614,* 1275–1283.

Menčík, J. (2016). Standards related to reliability. https://doi.org/10.5772/62366.

Moret, W. (2014). Vulnerability assessment methodologies: A review of the literature. A Report for USAID ASPIRES PEPFAR fhi360.

Southwick, S. M., Bonanno, G. A., Masten, A. S., Panter-Brick, C., & Yehuda, R. (2014). Resilience definitions, theory, and challenges: Interdisciplinary perspectives. *European Journal of Psychotraumatology, 5,* 25338.

Tanner, T., Bahadur, A., & Moench, M. (2017). *Challenges for resilience policy and practice.* London, UK: Overseas Development Institute.

www.ranker.com/list/worst-natural-disasters-2019/ranker-news.

Practical Concepts

4

Abstract

The discourse in this chapter focuses on concepts necessary for effective application of the value ethos, the sustainability paradigm, and the technical principles discussed in the preceding chapters. Curiously, these concepts tend to be overlooked or taken for granted; after all, the presumption is that asset management practitioners should be able to plan and organise relevant resources, schedule tasks and compose teams that can make informed decisions to ensure that tasks are executed effectively. The paradoxical effect of trivialising the concepts of planning, scheduling, organising, teaming and decision-making often vitiates the management of engineered assets, notwithstanding that there is widely available software that supposedly embed, formalise and automate these concepts.

4.1 Planning, Organising and Scheduling

As explained in the following sections, planning, organising and scheduling are indispensable practical concepts for managing an engineered asset. In fact, planning, organising and scheduling constitute the first practical activities in the management of assets.

4.1.1 Planning

Philosophically, planning derives from strategy and policy in that it involves anticipation of the future with a conscious determination of a course of action to

© The Author(s), under exclusive license to Springer Nature Switzerland AG 2021
J. E. Amadi-Echendu, *Managing Engineered Assets*,
https://doi.org/10.1007/978-3-030-76051-9_4

achieve desired results. Planning is a dynamic intellectual and mental process involving imagination, foresight and informed choice, and has been defined in a number of contexts as follows[1]:

- a process of developing strategies to achieve desired objectives;
- a process for establishing goals, policies, and procedures to achieve the goals;
- a process of setting goals, developing strategies, and outlining tasks and schedules to accomplish the goals;
- "the act of deciding how to do something".

Each definition implies that planning is a **process** which

i. focuses attention to a purpose or objective(s);
ii. is result-oriented, pervasive and rational;
iii. forms the basis for coordinating and organising resources;
iv. reduces uncertainty; and
v. stimulates innovation and creativity.

Referring to Chap. 2, an asset provides the means for the realisation of value, and management involves all activities necessary to ensure that an asset provides the means for the realisation of value. Since the management goal is to realise value from an asset, then planning not only involves thinking about how to realise value, but also what must be done to realise value. This means aligning and rationalising how to realise value against what must be done to realise value.

Planning is a primordial management function.[2] In terms of managing an asset, a plan must, at least, include:

i. a specified scope of activities necessary for the realisation of value from an asset;
ii. effort required, as expressed in terms of tasks that will be performed within the specified scope;
iii. identification and specification of all the resources that will be required and will be deployed to perform the identified tasks;
iv. a valuation (i.e., in monetary terms, a budget) encompassing (i), (ii), and (iii) to facilitate decision making with regard to the feasibility of the specified activities and tasks;
v. assumptions made regarding (i), (ii), (iii), and (iv) above; as well as
vi. a specification of the period of validity of the plan.

---

[1]Seely, J. R. (1962). What is Planning? Definition and strategy. *Journal of the American Institute of Planners, 28*(2), 91–97. https://doi.org/10.1080/01944366208979425.
[2]• Koontz, H., & O'Donnell, C. (1955). *Principles of management: An analysis of managerial functions.* New York: McGraw-Hill.
   • Jeseviciute-Ufartiene, L. (2014). Importance of planning in management development organisation. *Journal of Advanced Management Science, 2*(3).

4.1.2 Organising

Once a plan has been articulated as indicated in the preceding subsection, then the next logical step is to consider how to organise the resources that will be required to execute the plan. In a general sense, organising means systematic arrangement, alignment, integration and harmonisation of competences, data, information, knowledge, people, skills, tasks, material and non-material resources towards achieving the objectives specified in a plan. It is about ensuring that all the resources required to successfully execute a plan are in order and in place ready for deployment. Organising combines and synchronizes human, financial, tangible and intangible resources towards achieving the purpose of a plan. The process involves arranging these various types of resources into effective groupings so that the tasks identified in a plan can be successfully performed. Such arrangement establishes a hierarchy of relationships between competences, data, information, knowledge, people, skills, tasks material and non-material resources.

Taking into consideration the multidisciplinary nature of managing an asset, organising involves:

i. collation of competences, data, information, knowledge, material and non-material resources;
ii. grouping of activities and tasks;
iii. formation of people teams based on complementary and synergistic competences, knowledge and skills;
iv. linking data, information, knowledge, material and non-material resources to activities, tasks and teams;
v. assigning grouped activities and tasks to the respective teams;
vi. assigning each team with authority necessary for it to execute activities and tasks; and
vii. formalisation of the relationships between the teams so that they work together to perform all the activities and tasks required to ensure that value is realised from the asset.

4.1.3 Scheduling

A schedule states when something is supposed to happen; it shows the timing (start time and stop time) of activities and tasks.[3] Scheduling involves allocating human and material resources to appropriate teams and establishing the time period when the teams can successfully carry out the specified activities and tasks. A schedule specifies the activities and tasks that can be realistically performed by an appro-

---

[3]Baker, K. R. & Trietsch, D. (2009). *Principles of sequencing and scheduling*. Wiley.

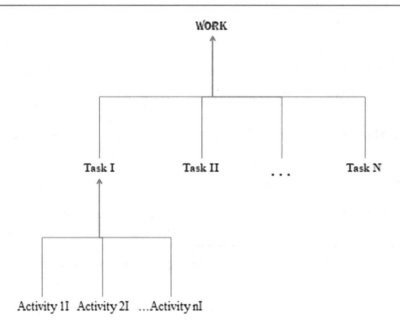

Fig. 4.1 A heirarchical decomposition of work into tasks and activities

priate team within the specified period of time given the available competences, data, information, knowledge, material and non-material resources.

The process of scheduling involves both the prioritisation of activities and tasks, and the optimisation of available capabilities[4] concomitant with the available time. A schedule must define:

   i. throughput—i.e., the amount of work (i.e., activities and tasks) that can be completed within a fixed period of time;

  ii. turnaround—i.e., the time required to complete an activity, task or work; and

 iii. timeliness—i.e., deadline for the completion of an activity, task, or work (see the hierarchical depiction in Fig. 4.1).

4.2   Team and Teaming

As indicated in the preceding sections, the work of managing engineered assets is a multidisciplinary endeavour predicated upon a wide range of knowledge areas, skills and competences. Such scope of capabilities can only be realised in the

---

[4]Capability = function (authority, competencies, data, information, knowledge, material and non-material resources, etc.).

context of a team. For this reason, the concepts of 'team' and 'teaming' are important for effective management of engineered assets.

4.2.1 Team

A team comprises a number of individual but complementary and synergistic competences, skills and resources (tangible and intangible). The purpose of a team is to perform and complete an activity, a task, a job, or a project. According to the BusinessDictionary,[5] "a team becomes more than just a collection of people when a strong sense of mutual commitment creates synergy, thus generating performance greater than the sum of the performance of its individual members". This definition implies that a team should naturally be composed of multiple disciplines required to succesfully execute a specified activity, a task or workpage. Thus, a group of employees in the same organisational unit (e.g., operations department, maintenance section, or finance department) may not necessarily constitute a team especially if the specifed activities, tasks, work package or problem to be solved demands a multi-disciplinary set of varying levels of knowledge, skills and competences.

In practice, team members are expected to:

- operate with a high degree of interdependence;
- share authority and responsibility for self-management;
- are accountable for the collective performance;
- work towards a common goal and shared reward(s).

Interestingly, as the era of Society 5.0 and 4IR technologies evolves, it is conceivable that asset management teams will comprise intelligent machines, humanoid robots and, of course, humans. With the envisaged advances in artificial intelligence, big data analytics and robotics, it may be that execution teams may be composed only of humanoid robots, particularly where assets are deployed in situations where human safety may be compromised.

4.2.2 Teaming

Teaming is the process of actively building and developing teams that can and must learn to deal with the vagarious requirements of managing engineered assets. Teaming is a "...dynamic activity, not a bounded, static entity" and "involves coordinating and collaborating without the benefit of stable team structures, because many operations like hospitals, power plants, and military installations require a level of staffing flexibility that makes stable team composition rare".[6] Learning is

---

[5]http://www.businessdictionary.com/definition/team.html.
[6]Edmondson, A. (2012). *Teaming: How organizations learn, innovate, and compete in the knowledge economy.* Wiley. www.josseybass.com.

the most important feature of teaming, so in addition to the complementarity of the multi-disciplines, each team member must not only be able to learn from both the internal and external environments but also, the learning must accentuate team performance towards achieving the desired objective.

In the era of Society 5.0 and 4IR technologies where it seems possible that teams could comprise intelligent machines, humanoid robots and humans, it is mind boggling as to what teaming would look like in the future! Curiously, humanoid robots are being used as personal assistants, for care giving and healthcare, education and entertainment, search and rescue, in manufacturing and maintenance, and particularly for research and space exploration. The increasing deployment of intelligent machines and humanoid robots will not only challenge conventional planning, organising and scheduling practices but also, represent a paradigm shift on the scoping of activities, grouping and prioritisation of tasks, and the notion of available time. The challenge will be exacerbated not only because of rapid evolution of big data and information required but also, by the fact that, planning, organising, scheduling and decision-making are continuous activities throughout the respective business cycles and life stages of an engineered asset.

4.3 Due Diligence

Due diligence[7] is essential for managing assets. The process of due diligence includes auditing, review and verification of all data and information captured and recorded against an asset. Although due diligence is generally regarded as part of investment appraisal during the acquisition of an asset, however, the process applies through all the stages in the life of an asset. In practice, a checklist is used and the investigative process of due diligence typically demands skill and competence at an expert level. Simply stated, due diligence involves thorough examination to establish the validity and accuracy of data and information used for decision making.

4.4 Decision Making

Making decisions is the essence of management.[8] It is a basic function of management and involves a process of choosing from a number of alternatives to achieve a desired outcome. Individuals, groups and teams make decisions, thus epistemological rules[9] inherently govern the process of decision making. Ideally,

---

[7]Bing, D. (1996). *Due diligence techniques and analysis: Critical questions for business decisions.* Quorum Books. ISBN 1-56720-029-X.

[8]Pušeljić, M., Skledar, A., & Pokupec, I. (2015). Decision-making as a management function. *Interdisciplinary Management Research, 11,* 234–244.

[9]Boulding, W., Moore, M. C., Staelin, R., Corfman, K. P., Dickson, P. R., Fitzsimons, G., Gupta, S., Lehmann, D. R., Mitchell, D. J., Urbany, J. E. & Weitz, B.A. (1994). Understanding managers' strategic decision-making process. *Marketing Letters, 5*(4), 413–426.

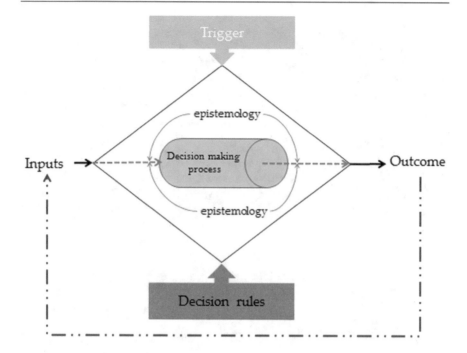

Fig. 4.2 A mapping of decision-making process

the process of making decisions should follow an explicit set of rules that serve to demonstrate coherency, consistency and validity,[10] albeit that the rules and associated logic may evolve to accommodate changing circumstances. This means that decision making can neither be arbitrary nor static. The epistemological dimension means that in practice, decision making often involves tacit rules (re: the cliché that the loudest voice gets heard!).

As illutrated in Fig. 4.2, the decision making process is usually triggered by events or a set of facts, for example, disruption to an asset, degradation in the performance or deterioration in the capability of an asset.

Also, as illustrated in Fig. 4.3, effective decision making must be informed, that is, based on factual and relevant evidence. Inputs to the decision making process include:

- data and information (e.g., benefits, costs, etc.);
- environmental factors (e.g., stakeholders' value perceptions and propositions);
- articulations of uncertainty in terms of risk ambiguity and opportunities;

---

[10]Haidar, A. D. (2016). Decision making principles. In *Construction program management—Decision making and optimization techniques* (pp. 25–55). Springer. https://doi.org/10.1007/978-3-319-20774-2_2.

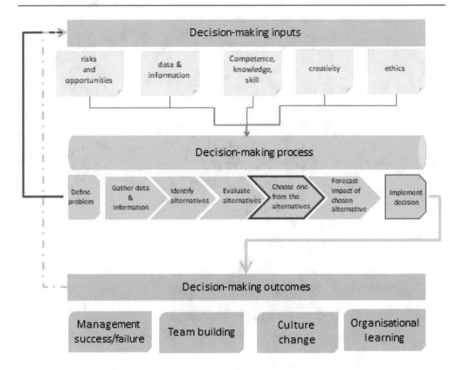

Fig. 4.3 Decision making process (adapted from Negulescu, O-H. (2014))

- epistemological factors—competency, creativity, ethics, knowledge, and skills of the decision maker(s); as well as,
- the outcomes of previous decisions.

The outcomes of the decision making process influence:

- management success or failure;
- organisational culture and learning[11];
- individual and/or team attitude and behaviour.

Whereas the imperative is that a decision, or a set of decisions must be made at a particular point in time, however, the challenge is to make appropriate decisions given the prevailing circumstances[12] and factors. For instance, three options to acquire an asset include (i) hire purchase, (ii) lease agreement, or (iii) outright ownership.

---

[11]Marchisotti, G. G., Almeida, R. L., & Domingos, M. L. C. (2018). Decision-making at the first management level: The interference of the organizational culture. *Revista de Administração Mackenzie, 19*(3). https://doi.org/10.1590/1678-6971/eRAMR180106a.
[12]McDonald, A. J., with Hansen, J. R. (2012). *Truth, lies and O-rings: Inside the space shuttle Challenger disaster*. James Hanson. ISBN 13: 9780813041933.

The decision imperative is to acquire an asset at a particular point in time but, the challenge is to choose the appropriate option given the prevailing circumstances and factors such as financial, market and technology risks. Another challenge is to integrate histories such as degradation models, cost models, *et cetera* to guide the decision to acquire an asset. Although decision-making is increasingly data and information driven, however, given the VUCA context of the era of Society 5.0, it is important that the epistemological dimension of the decision process is explicit so as to accentuate trust.

For brevity, decisions affect asset value and valuation *inter alia*, therefore, the stakeholders to an asset are interested in, and concerned about the process of decision making. An asset manager must make decisions regarding the acquisition and deployment, utilisation, and retirement of an asset. The reality is that decision making is continuous throughout the life of an asset.

4.5 Summary

The discourse in this chapter has reiterated that planning, organising, scheduling, teaming and decision making should be understood as extremely important concepts for effective management of assets. The presumption that these concepts are explicitly automated in widely available enterprise planning and decision making software embedded in ICT systems often lacks justification. The reality is that the knowhow to apply these concepts rely on the epistemological (e.g., bias, culture) plus the tacit and subjective dispositions of individuals, groups, and teams involved in the management of engineered assets.

4.6 Exercises

1. Using case studies, explain, in convincing detail, why the concepts discussed in this chapter are vital for managing engineered assets.
2. Choose an organisational unit or an organisation in a particular sector of human endeavour (e.g., agriculture and agro-allied, education and research, health, manufacturing, mining and minerals processing, security, transportation, utilities, etc.). Using specific examples, comment on whether any of the concepts discussed in this chapter is being applied by your chosen organisation.

References and Additional Reading

Baker, K. R. & Trietsch, D. (2009). *Principles of sequencing and scheduling*. Wiley.
Bing, D. (1996). *Due diligence techniques and analysis: Critical questions for business decisions*. Quorum Books. ISBN 1–56720-029-X.

Boulding, W., Moore, M. C., Staelin, R., Corfman, K. P., Dickson, P. R., Fitzsimons, G., et al. (1994). Understanding managers' strategic decision-making process. *Marketing Letters, 5*(4), 413–426.

Edmondson, A. (2012). *Teaming: How organizations learn, innovate, and compete in the knowledge economy*. Wiley. www.josseybass.com.

Haidar, A. D. (2016). Decision making principles. In *Construction program management—Decision making and optimization techniques* (pp. 25–55). Springer https://doi.org/10.1007/978-3-319-20774-2_2.

Jeseviciute-Ufartiene, L. (2014). Importance of planning in management development organisation. *Journal of Advanced Management Science, 2*(3).

Koontz, H., & O'Donnell, C. (1955). *Principles of management: An analysis of managerial functions* (p. 1955). New York: McGraw-Hill.

Marchisotti, G. G., Almeida, R. L., & Domingos, M. L. C. (2018). Decision-making at the first management level: The interference of the organizational culture. *Revista de Administração Mackenzie, 19*(3). https://doi.org/10.1590/1678-6971/eramr180106a.

McDonald, A. J., & Hansen, J. R. (2012). *Truth, lies and O-rings: Inside the space shuttle Challenger disaster*. James Hanson. ISBN 13: 9780813041933.

Negulescu, O.-H. (2014). Using a decision-making process model in strategic management. *Review of General Management, 19*(1), 111–123.

Pušeljić, M., Skledar, A., & Pokupec, I. (2015). Decision-making as a management function. *Interdisciplinary Management Research, 11*, 234–244.

Seely, J. R. (1962). What is planning? Definition and strategy. *Journal of the American Institute of Planners, 28*(2), 91–97. https://doi.org/10.1080/01944366208979425.

Engineering Asset Management Framework

5

Abstract

The discourse in this chapter introduces elements that constitute a framework for managing engineered assets. The discourse covers business and natural cycles during the life stages of an asset, the hierarchy and register of assets, as well as standards for the management of assets in the era of Society 5.0 concomitant with 4IR technologies.

5.1 Business Cycles, Natural Cycles and Asset Life Phases/Stages

There is generally a *laissez faire* understanding and application of the term asset life cycle, often resulting in confusing interpretation, measurement, reporting, and poor decision making. The following discourse is intended to provide some clarity with regard to the generally *laissez faire* obsfucation of business and natural cycles with asset life phases and stages.

5.1.1 Business and Natural Cycles

In order to achieve its goals, an established business organization such as an airline operator typically acquires, deploys and utilizes a wide range of engineered assets (e.g., aircraft, airport infrastructure and associated facilities) to provide air travel services to those wishing to travel by air. After a period of time, the business organization assesses and evaluates its progress, and reports its performance against the targets set for the period in question. For a formally registered business organisation, the content of the report and the interval between reporting time periods may be statutorily specified, and the reporting requirements may also be

J. E. Amadi-Echendu, *Managing Engineered Assets*,
https://doi.org/10.1007/978-3-030-76051-9_5

stipulated in company law, supporting legislation and attendant regulations. In this regard, the socio-economic and statutory obligations to report on the basis of specified time intervals establishes the formal business cycles, albeit that a business organisation may also adopt its own assessment periods separately from the formally specified reporting times.

Engineered assets that facilitate business activities exist in the world of the environment and ecology where the natural cycles of climate, weather and other *vis major* factors exert changing influences (stressors) on both the assets and business activities. The life of an asset (e.g., an aircraft) typically transcends several business and natural cycles. Hence, the life of an asset can be measured in terms of the number of business or natural cycles that the asset has actually been in existence. By convention, the life of an asset is generally stated either in hours, days or calendar months/years. However, it can be a misnomer to just state the life span of an asset in hours, days or calendar months/years of the asset's existence. The argument here is that the word 'cycle' aptly derives from the business and/or natural cycles. In principle, an asset has only one life, and the life does not cycle; albeit that in practice and according to the sustainability principle, the life of an asset may be 'extended', e.g., by replacing components that have failed or reached end-of-life.

5.1.2   The "Lifecycle" Argument

A non-repairable component of an asset can be replaced in order to sustain the same function when the component fails. Just like a failed non-repairable component, a spare part, subassembly, subsystem, or even the asset which no longer performs as stipulated may be decommissioned and replaced by another spare part, subassembly, subsystem, or asset which can perform the same function and deliver the desired levels of service, value or set of business objectives.

Depending on financial accounting rules, the cost of replacing the component, spare part, subassembly, or subsystem of an asset may or may not be capitalised. For a repairable item, the cost to repair and re-commission the item to perform and sustain the designated asset function(s) may be capitalised depending on prevailing financial accounting rules. Take for example, an asset such as an earth moving vehicle. A fuse is a non-repairable component of the asset and, expenditure on the replacement of a fuse is not likely to be capitalised. However, the engine of the earth moving vehicle could well be a repairable or replaceable subsystem, and depending on the financial accounting rules, expenditure on the replacement or repair and re-commissioning of the engine may be capitalised.

In essence, the replacement of a failed component, or the repair/replacement and re-commissioning of a spare part, subassembly or subsystem of an asset should restore the capacity of the asset to deliver the required level of service or value, and/or may extend the functional life of the asset. Take, for example, the re-chargeable battery in a mobile phone. The functionality of the phone can degrade to an unacceptable level when the battery has, say, only 3% of charge left. Recharging the battery to a mere 10% can restore full functionality of the mobile phone. The life of the phone (asset) has not expired, albeit that the battery (component) has been recharged to restore full functionality of the phone. As implicit in rechargeable battery design, the fact is that the useful life of the battery can be extended by recharging but, this does not mean that the life of the battery has cycled. The life of the battery ends if and when it cannot be re-charged, and the life of the re-chargeable battery is not the same as the life of the mobile phone, so it may not be prudent to replace the phone on expiry of the life of the battery unless the battery is not a detachable component or spare part of the phone. A similar argument holds in terms of the relationship between an asset and its components, subassemblies, or subsystems.

5.1.3 Asset Life Phases and Stages

The life span of an asset may be resolved into a number of phases and stages irrespective of how the asset life is stated. In this book, the interpretation is that a phase in the life of an asset comprises at least one business cycle, e.g., commissioning or de-commissioning phase; and one or more phases constitute a life stage. There are at least three viewpoints that can be applied to describe the life phases and stages of an asset.

(a) *Technology Development/Product lifecycle management (PLM [1]) perspective*

The first viewpoint is the technology development or so called product lifecycle management (PLM) perspective depicted in Fig. 5.1. This perspective more or less derives from the fact that an engineered asset is a physical and tangible form of technology[2], so the life phases transcend from conception, through development and prototyping, to manufacturing or production. The product (i.e., asset) may then be acquired by a user while the manufacturer/producer, vendor or supplier provides technical support that may include replacements and technology upgrades of components and spare parts that have short life spans.

The circular PLM model depicts the fact that accumulated knowhow can be 'recycled' to manufacture or produce a new version of the same product; in essence, the old product ceases to exist. Alternatively, failed components of the old product (i.e., defunct asset) may be replaced or obsolete parts upgraded to restore functionality at the desired level of performance. The recycling of accumulated knowhow coupled with the replacement or upgrading of some components may result in a technologically superior product (asset), even though the asset performs the same function in practice.

Inadvertently, the term 'lifecycle' has more or less prevailed from the historical antecedents of PLM, MRO, and MRP. At least, two perplexities arise from the

---

[1]Terzi, S., Bouras, A., Dutta, D., & Garetti, M. (2020). Product lifecycle management-From its history to its new role. *International Journal of Product Lifecycle Management 4*(4):360–389. https://doi.org/10.1504/ijplm.2010.036489.
[2]Technology can exist as (i) a method/process/technique, (ii) knowledge/knowhow, (iii) product/service, or (iv) combination of (i), (ii), and (iii).

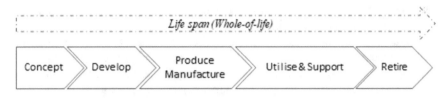

❖ *Technology Development viewpoint: PLM phases*

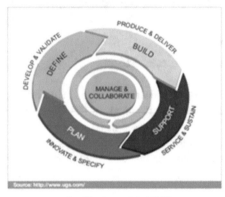

❖ *Circular model of PLM*

Fig. 5.1 Technology development life stages

incongruent use of the term lifecycle. The first is the fact that a component, spare part, subassembly, or subsystem of an asset can be replaced so as to restore, or even to improve functional performance of the asset. Such restoration of, or improvement in functional performance is often technically construed as life extension, i.e., giving the asset a new 'lease on life'. Although the replacement of a component or the reuse of a repaired spare part may result in an extension of the useful life of an asset, however, it does not change the fact that each component, spare part, subassembly, or subsystem of an asset, and even the asset itself, really has only one life.

The second and very puzzling feature of adopting the circular model of PLM (Fig. 5.1) is the suggestion that 'planning' is a phase in the life of the product (asset). As we shall discuss later, from the asset management viewpoint and practice, the reality is that planning happens throughout the span of life of an asset.

(b) *Technology Implementation/Project Management perspective*

The second viewpoint emanates from the technology implementation or project management perspective (see Fig. 5.2). This viewpoint combines systems engineering and project management perspectives in the sense that an asset transcends from the concept and feasibility phases, through the design, construction, installation and commissioning phases, to the asset deployment and utilisation, then retirement (termination and decommissioning).

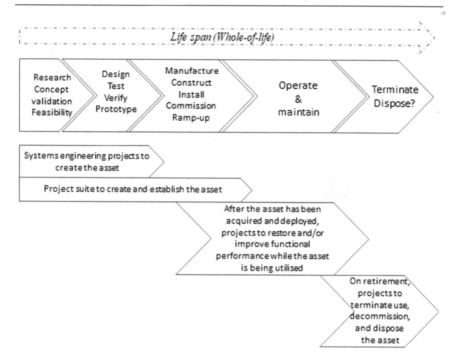

Fig. 5.2 Technology implementation life stages

The life of a project could begin and end within a phase, for example, the construction phase could be defined as a project. Once an asset is deployed to perform a function and during the utilisation phase, several projects may be carried out to, for example, replace a component, repair a subassembly, modify or upgrade a subsystem, or even to terminate and decommission the asset as situations dictate.

Every project that happens throughout the life of the asset must be planned and executed. While the required function persists, there may be several replacement projects planned and executed during the utilisation stage in the life of the asset. The accumulated knowhow can be reused or recycled to improve project performance. Thus, the term 'lifecycle' can be aptly applied to the reuse/recycling of intellectual capital and accumulated knowhow related to the projects. Even if the reuse/recycling of knowhow improves performance of the project and leads to an extension of the useful life of the asset, however, the fact remains that the life of each failed non-repairable component, failed or malfunctioning subassembly, subsystem, or asset ends once the component, spare part, subassembly, subsystem, or even the asset itself is removed and replaced by another.

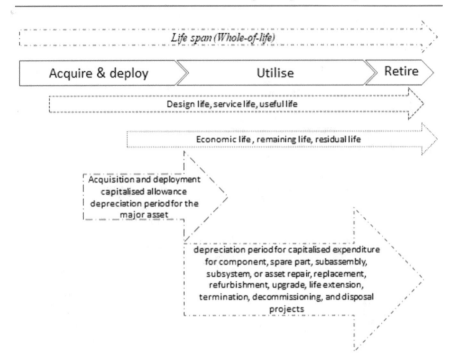

Fig. 5.3 Asset life stages

(c) *Technology Deployment/Asset Management perspective*

The third and pertinent perspective is the technology deployment or asset management context. In this context, the life of an asset can be surmised into three pragmatic stages, viz: (i) acquisition and deployment, (ii) utilisation, and (iii) retirement (see Fig. 5.3). Each life stage can be further resolved into a number of phases, for example, the acquisition stage incorporates the first three phases depicted in technology development and implementation contexts (a) and (b) above. For the user, the asset life commences when an asset is acquired and deployed for particular function(s) until when the utilisation is terminated, and the asset is retired, i.e., decommissioned from performing the designated function(s).

5.1.4   Perceptions of Asset Life

The three perspectives also give rise to various definitions of asset life such as design life, economic life, useful life, service life, remaining life, and residual life (see also Chap. 8). The definitions and applications of asset life are often subjective to the persuasions of the various disciplines involved in managing engineered assets. For example, engineers and technicians tend to be more conversant with design life and remaining life. Accountants tend to be more concerned about economic life in particular. It is a common misnomer to equate asset life as the period specified for capital allowance depreciation (i.e., when the capital expenditure has been written off, or the capitalised value of the asset has been fully amortised in accordance to prevailing accounting rules). In practice, the accounting depreciation period depends more on financial capital taxation rules and less on asset life considerations.

Irrespective of the asset life considerations (see Fig. 5.4), a business organisation typically acquires and deploys an asset, utilises the asset for the specified or desired purpose, and then chooses to retire the asset when the asset

(i)   is no longer functioning as intended, and/or
(ii)   is no longer capable of delivering the level of service or performance required, or
(iii) destroys inherent value of the asset and the business, or
(iv)   does not provide the means for creating value, or
(v)   does not lead to the realization of the investment goals, or
(vi)   adversely affects the set of business objectives that may be defined in terms of (i)–(v) above.

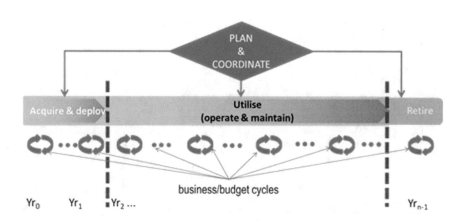

❑ the life of an asset (e.g., an aircraft) transcends several business cycles

❑ the span of life of an asset can be measured either in terms of

   • calendar days/months/years, or more appropriately,

   • as the number of business cycles that the asset has actually been deployed and utilised

Fig. 5.4 Asset life considerations

5.2   Influence of Preceding Decisions on Asset Life Phase/Stage

The asset management viewpoint adopted throughout this book makes practical sense, particularly because the ramifications of the choices (i.e., decisions) made in preceding stage(s) manifest in subsequent life stages. For instance, the strategy adopted and implemented for acquiring and deploying the asset can impose profound consequences during utilisation and retirement stages of the life of the asset. Furthermore, how the asset is utilised will invariably influence the condition in which the asset will be retired. The picture in Fig. 5.5 is an illustration of the predominant influence of the acquisition and deployment phases/stage on the latter stages of the life span of an asset. Anecdotal and empirical evidence indicates that choices made during the acquisition and deployment stage can have up to a 95% influence on the latter stages of the life span of an asset. This implies that decisions that can be made during the utilisation and retirement stages are consequent upon the choices already made when the asset was acquired and deployed.

Obviously, decisions made during the utilisation stage will increasingly influence decisions that can be made when the asset has to be retired. The corollary is that choices made during earlier phases/stages can and do constrain choices and decisions that can be made during latter phases/stages of the life of an asset. From the preceding discourse, it is vital to discern the asset structure, that is, whether

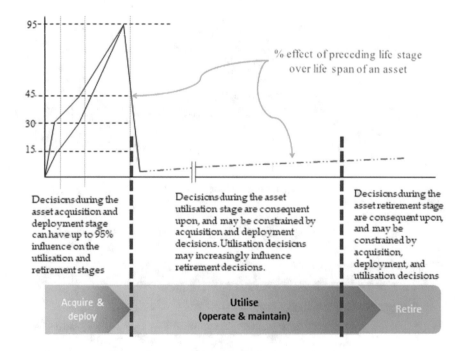

Fig. 5.5 Influence of preceding phases/stages decisions on the life span of an asset

asset life is being considered at the component, spare part, subassembly, or subsystem of the asset hierarchy.

5.3 Asset Structure

By conceptualisation and design, every engineered asset has a structure. The structure is more or less inherent to the asset, and determines how and what data and information can and should be captured and recorded against the asset. Fundamentally, the configuration of the asset structure (that is, the asset hierarchy) in conjunction with the corresponding registration of appropriate data and information provides the means to regulate how an asset is managed. Therefore, the unambiguous configuration of the asset hierarchy together with the establishment of a detailed asset register is a primordial requirement for the management of engineered assets.

5.3.1 Asset Hierarchy

Figure 5.6 illustrates a generic hierarchical structure of an asset. The hierarchy depicts both a *bottoms up* aggregation of the composition of the asset, or,

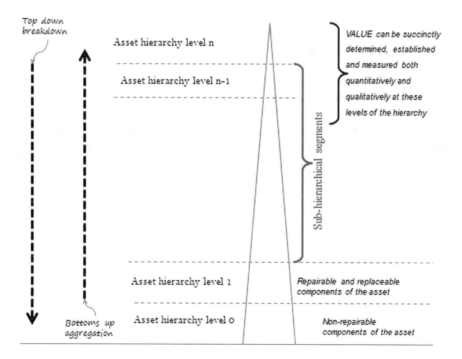

Fig. 5.6 A conceptual asset hierarchy

alternatively, a *top down* resolution of the components of the asset. The hierarchy may be configured in terms of (i) the spatial arrangement or physical layout of the asset, and/or (ii) the financial rules for recognising assets, and/or (iii) the process engineering composition of the asset (in terms of energy flow patterns), and/or (iv) the specific and respective functions of the components of the asset, or (v) as combinations of (i), (ii), (iii) and (iv).

From the definition of an asset, it is important to state what the *value* is at each level of the hierarchy. Ideally, *value* must be succinctly determinable and measurable both quantitatively and qualitatively at each level of the hierarchy. At higher levels of the hierarchy, it may be easier to define and state *value* broadly in terms of return on investment (i.e., value-add), effectiveness, compliance, reputation, etc. At the lowest level of the hierarchy, it is much easier to simply state economic cost, qualified by the risk of failure of the replaceable and/or non-repairable components of the asset. For instance, if an articulated backhoe loader or self-loading concrete mixer is defined as a discrete asset, then the engine is a replaceable or repairable subsystem of such an asset whereas the engine gasket is a replaceable but non-repairable component of the asset. The number of segments between the highest and lowest levels of the asset structure depends on what is defined as the asset and how the hierarchy is consequently configured.

5.3.2 Asset Register

Different types of data and information concerning the asset may be captured and recorded at the appropriate levels of the hierarchy (see Fig. 5.7). Some of the data and information may be fixed records such as acquisition date, cost and supplier. Others may be changing transactional data and information such as operational cost, condition, failure rate and remaining life. In fact, it is relatively easy to capture economic cost at all levels of the hierarchy but, determining and measuring *value* succinctly at the lowest levels of the hierarchy may not be trivial. Capturing and recording data and information according to the asset hierarchical structure facilitates appropriate assembly, collection, clustering, and collation of data and information for reporting purposes. Such structured aggregation of data and information also facilitates informed decision making by the relevant functionaries within the organisation or entity responsible for managing an asset or a system of assets.

In essence, an asset register is a record of data and information about the components, spare parts, subassemblies and subsystems constituting an asset. The register specifically contains data and information captured and recorded according to the hierarchical structure of the asset. The register is a consolidated mapping of all data and information captured and recorded according to the hierarchical structure of the asset for the purposes of tracking, decision making and control, and reporting. Strictly speaking, an entity should have only one asset register. This is

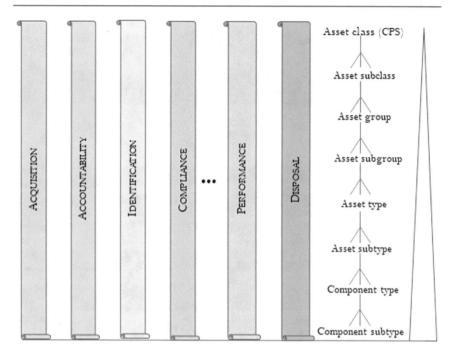

Fig. 5.7 Asset hierarchy: data and information categories

irrespective of the fact that the listing of the assets with the corresponding data and information may be concatenated into several databases and computerised information systems.

As illustrated in Fig. 5.8, different types of data captured and recorded against an asset may be grouped into categories that make reporting easy to comprehend during the respective stages in the life of an asset. For example, data and information about the condition of an asset, the date of termination for use or decommissioning, and the asset's residual value are pertinent during the retirement stage of life, and may be conveniently grouped as "DISPOSAL" data and information. Thus, capturing, collating, recording, and reporting of data and information are continuous asset management activities. For compliance purposes, the grouping of data and information is subject to guidelines, regulations and standards.

To avoid confusion, it is recommended to construe the asset hierarchy independently from the organisational structure. Such distinction ensures that various data and information concerning the asset are recorded at the appropriate level of the hierarchy. This, in turn, facilitates appropriate decision making by the relevant functions within the organisation structure.

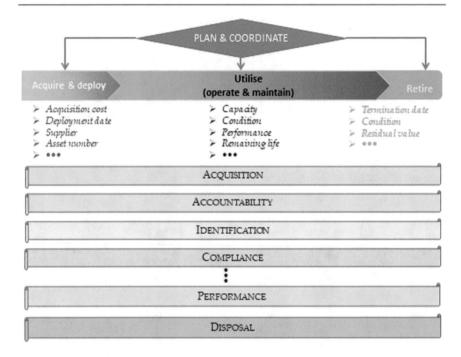

Fig. 5.8 Asset life stages: data and information

5.3.3 Influence of 4IR Technologies

4IR technologies such as augmented reality, distributed ledgers, and internet-of-things (IoT) can be applied as depicted in Fig. 5.9. These technologies can be used for identification, cataloguing, inventory control, creation of immutable records, facilitation of transparent data and information flows, digital twinning,[3] *et cetera*. For instance, data and information at the various levels of the asset hierarchy can be encrypted in a blockchain (i.e., recorded in a way that makes it difficult or impossible to change the recorded data and information). Conceptually, each component, spare part, subassembly, subsystem, or even the whole asset can be IoT-enabled to facilitate connectivity and the exchange of data with other devices over the internet. The cliché is that everything can be made 'smart.'

In the context of digitalisation, an asset is anything that has inbuilt connectivity (e.g., components, spare parts, subassemblies, subsystems, as well as materials, engineering drawings and wiring diagrams, and even contracts and purchase/sales orders). This implies that every engineered asset can have a digital twin. Digital twining facilitates connectivity and communicability between devices, systems, and

---

[3]Macchi, M., Roda, I., Negri, E., Fumagalli, L. (2018). Exploring the role of digital twin for asset lifecycle management. *IFAC PapersOnLine 51–11*, 790–795.

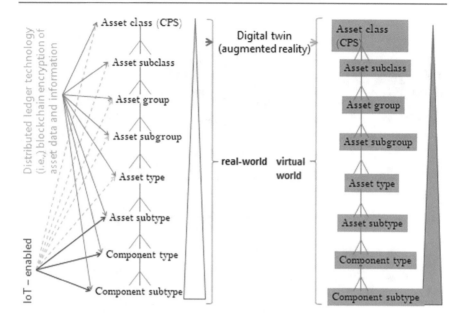

Fig. 5.9 Asset hierarchy and 4IR technologies

services within a network. This enables the identification of every asset, the asset's type, operational and technical data, status, and other asset-specific information.

It is in this regard that the reference architecture model for Industrie 4.0 (RAMI 4.0[4]) (see pictures in Fig. 5.10) provides a useful framework for the application of 4IR technologies to facilitate the management of assets in the era of Society 5.0. "RAMI 4.0 defines a service-oriented architecture where application components provide services to the other components through a communication protocol over a network." The reference architecture model is a three-dimensional digitalised platform that builds upon the networking strengths of the 7-layer open systems interconnection (OSI) model. To address all aspects involved in the management of industrial asset systems, the RAMI 4.0 includes an asset administration shell (AAS) as the interface that connects an asset to the network and three-dimensional digitalised platform. The AAS is designed not only to serve as the standardised communication interface but also, it is designed to contain all data and information captured, stored and recorded about an asset. Hence, an asset can have its own

---

[4]Sony, M. (2020) Pros and cons of implementing industry 4.0 for the organizations: A review and synthesis of evidence. *Production & Manufacturing Research. 8*(1).

Alcácer V., & Cruz-Machado V. (2019). Scanning the industry 4.0: A literature review on technologies for manufacturing systems. *International Journal of Engineering Science and Technology, 22*(3), 899–919.

Pisching, M. A., Pessoa, M. A. O., Junqueira, F., dos Santos Filho, D. J., & Miyagi, P. E. (2018). An architecture based on RAMI 4.0 to discover equipment to process operations required by products. *Computers and Industrial Engineering, 125,* 574–591.

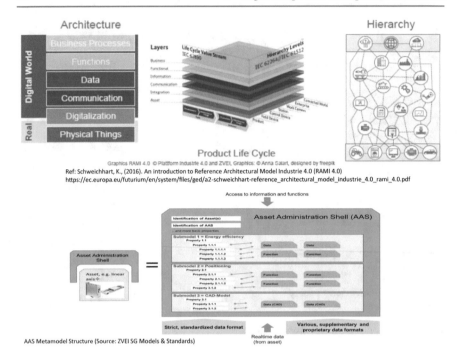

Ref: Schweichhart, K., (2016). An introduction to Reference Architectural Model Industrie 4.0 (RAMI 4.0)
https://ec.europa.eu/futurium/en/system/files/ged/a2-schweichhart-reference_architectural_model_industrie_4.0_rami_4.0.pdf

AAS Metamodel Structure (Source: ZVEI SG Models & Standards)

Fig. 5.10 RAMI 4.0 and the asset administration shell

administration shell even if an asset is defined as a discrete device (e.g., a sensor or a machine), otherwise, "several assets can form a thematic unit with a common administration shell." Furthermore, in a manner that mirrors the asset hierarchy, the combination of several thematic units can also have a common administration shell. With this approach, there is no doubt that RAMI 4.0 will spur further development of other guidelines and standards.

5.4   Asset Management Guidelines, Regulations and Standards

The picture in Fig. 5.11 illustrates that there is a myriad of guidelines, regulations and standards applicable towards the management of engineered assets. For brevity, the standards may be broadly categorised into suites such as—accounting and finance; environmental; engineering, technical, and technological; and management systems. In general, *legislative or statutory standards*, as well as regulations and safety standards tend to be mandatory and require full compliance. Notwithstanding the categorisation, standards may be subject to legislation and the level of compliance may also be jurisdictional.

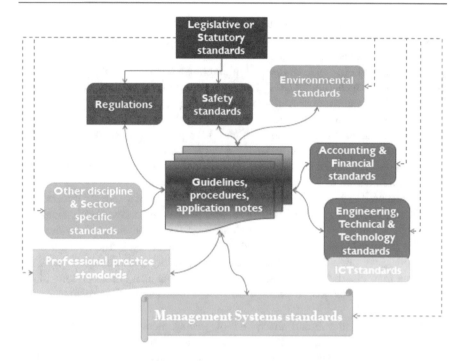

Fig. 5.11 AM guidelines, regulations and standards

Environmental standards emanate from the fact that all human activities generate pollution and waste. Thus, environmental standards are intended to "regulate and reduce the amounts and/or complexity of wastes discharged to the environment, with the ultimate objective of achieving"[5] minimal impact on ecology. Since all human endeavour is facilitated by engineered assets, consequently, environmental standards also regulate how engineered assets are managed.

Accounting and financial standards are sets of common rules for consistent, transparent and comparable maintenance of accounting records and the reporting of financial statements. It is essential that data and information pertinent for accounting and financial reporting are captured and recorded, especially at the appropriate levels of the asset hierarchy so as to facilitate proper decision making.

Engineering, technical, and technology standards encompass extensive suites of standards that may be sub-categorised in terms of disciplines (e.g., computing, civil, electrical/electronic, mechanical, information technology). The standards include, for example, data sheets, drawings, guidelines, procedures, requirements specifications, et cetera about the design, construction, installation, commissioning, operation, maintenance, replacement, upgrade and disposal of components, spare

---

[5]Bhaskar Nath Environmental Regulations and Standard Setting—Encyclopedia of Life Support Systems (EOLSS) http://www.eolss.net..

parts, subassemblies, and subsystems of an asset or the entire asset. Application notes generally provide very specific details about the technology or technologies embodied or inherent in the asset.

As depicted in Fig. 5.11, it is worth highlighting a special suite of information and communications technologies (ICT) standards, especially those that specify protocols for connectivity and communication between devices and networks that facilitate the control, operation and maintenance of discrete and systems of assets. These standards have evolved rapidly from 4 – 20 mA and 3–15 psi signals, RS-(232, 422, 485), fieldbus, and industrial ethernet and networking protocols to 4IR digitilized platforms (re: RAMI 4.0). Incidentally, there are many extant cyber physical systems that still deploy combinations of these generations of ICT protocols.

As illustrated in Fig. 5.11, each category/sub-category of standards may contain a variety of application notes, guidelines, and procedures necessary for managing an engineered asset. A very interesting categorisation is the suite of *professional practice standards* typically developed and/or promoted by learned societies and practitioner institutions, e.g., municipal engineers, facility asset managers, etc. Such standards may also be encapsulated in legislation, thus making them compulsory for professional practice, especially for the certification of particular brands of practitioners.

Progression in the broad discipline of management has also fostered the grouping of sub/categories of guidelines, regulations and standards into function-specific management systems. Furthermore, the proliferation of function-specific management practices has given rise to a suite of standards categorised as *management systems standards*. A justification for these management systems standards is that they facilitate integration and harmonisation whilst concurrently reducing inefficiencies in management practices. Another justification for management systems standards is that they can be used as blue-prints for auditing, benchmarking and certification of management practices. Notable among these suites of management systems standards are the ISO standards, and particularly, the ISO 5500× series. Interested readers are welcome to engage the significant amount of extant literature on ISO 5500× series of management systems standards for asset management.

5.5   The Asset Management System

The asset management system as illustrated in Fig. 5.12 arises as a consequence of the integration of function-specific asset management subsystems. The function-specific subsystems may be integrated and configured to serve each of the life stages of the asset, however, it is conceivable that most systems will transcend through the whole life of an asset. Thus, such systems should be implemented to cater for all the life stages of asset management. Ideally, the systems used for acquiring an asset should be transformed to facilitate the management processes during both the utilisation and retirement stages.

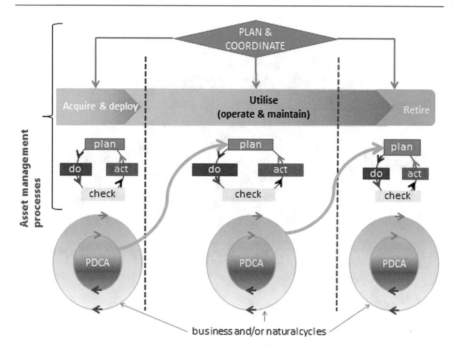

Fig. 5.12 Asset management system

The picture in Fig. 5.11 depicts that the plan-do-check-act (PDCA) loop is the core of the asset management system. This loop is the management kernel and the catalyst for continuous improvement. The asset management activities subsumed within the PDCA loop should be continuously performed within the business and/or natural cycles that occur during each of the phases that constitute the life stages of an asset. Thus, as illustrated in Fig. 5.12, the PDCA loop together with the business and natural cycles form concentric management circles through the phases and stages in the life of an asset.

5.6   Asset Management Framework

In the simplest sense, individuals, public and private entities, groups, and organisations invest in engineered assets in order to obtain benefit. As indicated earlier in Chap. 2, the paramount goal of asset management is to ensure that the investment ($\check{v}$) is turned into value v. An overarching framework for managing engineered assets is depicted in Fig. 5.13. This picture shows that people, organisation structure, and processes constitute the triple helix requirements necessary for managing engineered assets.

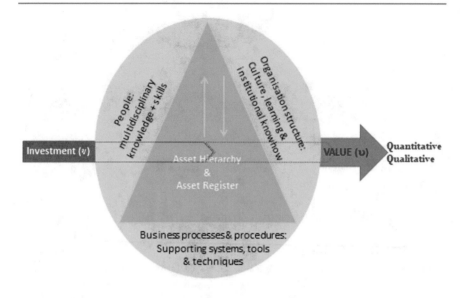

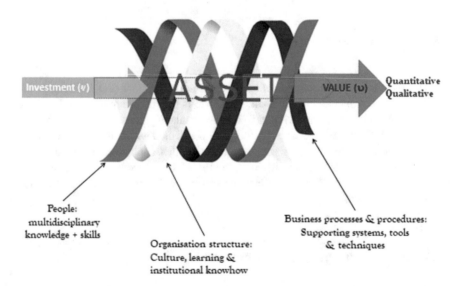

Fig. 5.13 Asset management framework

5.6.1   People Manage Engineered Assets

The multidisciplinary knowledge, knowhow, skills and competences for managing assets are retained and provided by human beings. As discussed earlier in Chap. 4, the concept of teaming requires that these human resources, that is, the multidisciplinary knowledge, knowhow, skills and competences embedded in humans are configured into teams that can direct and efficiently apply the non-human resources to ensure that the assets are properly managed. Effective application of the human resources is influenced by (i) how the teams are aligned, (ii) how the organisation is structured, as well as (iii) how well the processes are designed and followed.

5.6.2   Organisational Structure

Incidentally, organisations are typically structured in terms of conventional or traditional disciplines but not necessarily in terms of teams. This convention has often manifested in the so called 'silo' mentality and culture that tends to impair the behaviour, functioning and performance of asset management teams.

5.6.3   Processes and Procedures

The third helix is processes (what to do and how to do); they must be appropriately designed and implemented to ensure that asset management is effective. Procedures are part and parcel of processes and constitute the basis for the myriad of guidelines, regulations and standards applicable for the management of engineered assets.

5.7   Summary of the Chapter

The life of an engineered asset transcends many business and natural cycles through the stages of acquisition and deployment, utilisation (i.e., operations and maintenance), and retirement. The decisions made during the acquisition phases impose profound influences that manifest during the utilisation and retirement stages in the life of an asset. The asset hierarchy and register provide the primordial skeletal framework for managing assets. 4IR technologies, in particular the RAMI 4.0, will greatly enhance connectivity and the exchange of data between IoT-enabled components, spare parts, subassemblies and subsystems of assets as well as between systems of assets. Although a myriad of standards is necessary for the management of assets, however, *management systems standards* are providing the basis for increased formalisation and conformity in the practice of asset management. People, organisational structure, and business process systems interact in a triple helix manner to ensure that assets are effectively managed.

5.8   Exercises

1. Should an organisation have more than one asset register? Explain and justify your answer.
2. Suppose that you have been appointed to provide consulting advice to an organisation in any sector. Review some case studies, and then advise your client whether or not auditing and certification according to the ISO 55000/1/2 standards will result in improved asset management practice by the organisation.
3. Suppose that you have been appointed to provide consulting advice to an organisation in any sector. Review some case studies, and then advise your client on how 4IR technologies may elevate asset management practice within the organisation.

References and Additional Reading

Alcácer V., & Cruz-Machado V. (2019). Scanning the industry 4.0: A literature review on technologies for manufacturing systems. *International Journal of Engineering Science and Technology, 22*(3), 899–919

Bhaskar Nath Environmental Regulations and Standard Setting—Encyclopedia of Life Support Systems (EOLSS) http://www.eolss.net

Kulkarni, K., Kulkarni, V. N., Gaitonde, V.N., & Kotturshettar, B.B. (2021). State of the art review on implementation of product lifecycle management in manufacturing and service industries. In *AIP conference proceedings.* Vol 2316. pp. 030012. https://doi.org/10.1063/5.0036547

Macchi, M., Roda, I., Negri, E., & Fumagalli, L. (2018). Exploring the role of digital twin for asset lifecycle management. *IFAC PapersOnLine, 51–11*(2018), 790–795.

Pisching, M. A., Pessoa, M. A. O., Junqueira, F., dos Santos Filho, D. J., & Miyagi, P. E. (2018). An architecture based on RAMI 4.0 to discover equipment to process operations required by products. *Computers and Industrial Engineering, 125*, 574–591

Schweichhart, K. (2016). An introduction to reference architectural model industrie 4.0 (RAMI 4.0). https://ec.europa.eu/futurium/en/system/files/ged/a2-schweichhart-reference_architectural_model_industrie_4.0_rami_4.0.pdf

Sony, M. (2020) Pros and cons of implementing industry 4.0 for the organizations: A review and synthesis of evidence. *Production & Manufacturing Research. 8*(1)

Terzi, S., Bouras, A., Dutta, D., & Garetti, M. (2020). Product lifecycle management-From its history to its new role. *International Journal of Product Lifecycle Management 4*(4):360–389. https://doi.org/10.1504/ijplm.2010.036489

Asset Acquisition Stage

6

Abstract

The discourse in this chapter highlights issues to be considered when acquiring assets. Although the issues are not unique to the acquisition stage, however, their significance during the acquisition stage is remarkable, especially because, the effects of choices and decisions made during the acquisition stage manifest as challenges to be overcome through the subsequent life stages of an asset. In the parlance of practitioners, the acquisition stage in the life of an asset is often referred to as a 'capital development project', therefore, the considerations involve similar issues. For instance, the types of risks encountered in capital development projects also manifest during the acquisition and deployment stage of an asset, and some of the risks particularly impact on readiness to utilise an asset.

6.1 Asset Acquisition Considerations

The options for acquiring an asset depend on a number of considerations broadly grouped into funding and financing, technology, socio-economics and socio-politics, as well as issues that pertain to the natural environment and ecology. The issues are not uncorrelated, thus, the respective groupings of the issues may not be exclusively considered when an option to acquire an asset is being decided.

6.1.1   Funding and Financing Issues

The existence of extensive discourse[1] on project finance points to the fact that funding and financing are paramount challenges to the management of engineered assets. Issues such as liquidity, cost of capital, capital structure, risk/return, *et cetera*, exert prominent influence on the option eventually chosen to acquire an asset. In circumstances where funds are obtained through financing, a paramount concern is how macroeconomic factors[2] influence the cost of capital. The obvious questions include where and how to obtain funds, how to allocate funds, and how to expend funds to accomplish the goals of acquiring an asset. Detailed answers[3] to these and associated questions may be extrapolated from extant literature on funding and financing of capital projects.

Capital allowance is an important funding and financing consideration during the acquisition stage in the life of an asset. The allowance is typically a tax relief applicable to a qualifying expenditure such as an engineered asset. It informs accounting depreciation of the cost of acquiring an asset.[4] Accounting depreciation refers to writing-off the acquisition cost of the asset over time according to pre-scribed rules (re: accounting standards). In most jurisdictions, the acquisition cost of industrial plant and facilities generally qualifies as capital allowance expenditure that can be written-off.

The rules for financial recognition of assets generally specify qualifying expenditure for capital allowance[5] purposes. Paradoxically, the financial recognition of a component, spare part, subassembly, subsystem, or even an asset may not necessarily align with how an asset hierarchy is configured (see Fig. 6.1). This apparent misalignment often creates conundrum among the disciplines involved in managing engineered assets. The conundrum typically manifests during the utilisation and retirement stages in the life of an asset. Financial recognition rules more or less determine whether or not the cost to replace a component, spare part,

---

[1]Müllner, J. (2017). International project finance: Review and implications for international finance and international business. *Manag Rev Q, 67*, 97–133. https://doi.org/10.1007/s11301-017-0125-3.

Kayser, D. (2013). Recent research in project finance a commented bibliography. *Procedia Computer Science, 17*, 729–736.

[2]Rao, V. (2018). An empirical analysis of the factors that influence infrastructure project financing by banks in select asian economies. Asian Development Bank Economics Working Paper No.554.

[3]Walter, I. (ed.) (2016). In *The infrastructure finance challenge*. Cambridge, UK: Open Book Publishers. http://dx.doi.org/10.11647/OBP.0106.

IATA. (2016). Airline Disclosure Guide: Aircraft acquisition cost and depreciation. International Air Transport Association/KPMG.

Dugdale, D., & Abdel-Kader, M. (1999). funding issues in a major strategic project: a case of investment appraisal. *Accounting Education: An International Journal, 8*(1), 31–45. ISSN 0963-9284. http://dx.doi.org/10.1080/096392899331026.

[4]Yussof, S. H., Isa, K., & Mohdali, R. (2014). An analysis of the gap between accounting depreciation and tax capital allowance in Malaysia. *Procedia—Social and Behavioral Sciences, 164*, 351–357.

[5]Examples on calculations of capital allowances may be found in textbooks on tax and financial accounting.

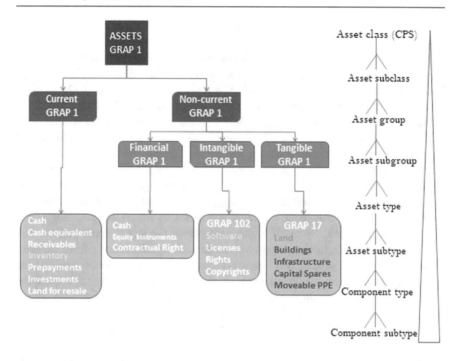

Fig. 6.1 Capital expenditure and asset structure

subassembly, subsystem is deemed to be capital or operational expenditure, even though concerns about reliability and criticality may inform persuasive arguments to capitalise the cost of acquiring an item.

It can be a misnomer to assume that accounting depreciation due to capital allowance affects the financial valuation of an asset. The financial valuation of an asset takes into consideration the condition of an asset. Valuation considers that the performance of an asset will generally degrade or deteriorate with time, especially due to aging and usage, or the combined effect of both phenomena (albeit that ageing is influenced by usage and vice versa). In practice, a comprehensive assessment of condition usually informs the decision to retire an asset, despite the fact that the cost of acquiring the asset may have since been written-off, tax wise. The period for accounting depreciation of the capital cost of an asset is generally a prescribed norm, whereas the actual life of an asset may be shorter or longer than the prescribed accounting depreciation period. It is important to distinguish between the time allowed for the capital expenditure to be tax written-off and the actual duration of life of an asset. The distinction often has implications on how remaining life, useful life, residual value and salvage value are defined and interpreted across sectors and industries, and also, by the various disciplines involved in the management of engineered assets. Inadvertently, technology obsolescence significantly influences the determination of the life of an asset.

6.1.2   Technology and Technological Issues

An engineered asset represents explicit corporeal embodiment of technology in its various forms, given that technology can exist in tangible and/or intangible forms of knowledge, process, and physical artefact. A highly significant issue with regard to asset management is the particular risk of technology obsolescence, given the prevalence of rapid technological innovations as the era of Society 5.0 evolves. This risk is magnified by the fact that the components, spare parts, subassemblies, and subsystems of an asset, and the asset itself have varying spans of life. Often, the variation in the spans of life of components, spare parts, subassemblies, and subsystems are exacerbated by technological innovation[6] plus other intervening factors. For instance, a manufacturer may be forced to upgrade a spare part of an asset in order to take advantage of the latest evolution in materials technology. The upgrade will automatically result in obsolescence of the legacy spare part.

The risks due to technological evolutions demand consideration during the asset acquisition stage for at least two reasons. Firstly, technological innovation and the rate at which obsolescence occurs influence the choice as to whether to acquire the asset by outright ownership, hiring or leasing, or any legal form of partnership arrangement. Notwithstanding the option chosen to acquire an asset acquisition, it is vital to appropriately identify and allocate the risks due to technological evolution. Secondly, technology obsolescence especially influences later decisions whether to repair, replace, upgrade or dispose legacy components, spare parts, subassemblies, and subsystems of an asset, and even the entire asset.

6.1.3   Environment and Ecology Issues

The following quotation reiterates the significance of environmental considerations as a requirement for managing engineered assets. "Institutional investors are increasingly committed to integrating environmental, social and governance (ESG) factors into their strategies for delivering risk-adjusted returns and delivering their ownership responsibilities.[7]" The imperative to transition towards

---

[6]Sandborn, P. (2007). In G. Bayraksan, W. Lin, Y. Son, & R. Wysk (Eds.), *Designing for Technology Obsolescence Management Proceedings of the 2007 Industrial Engineering Research Conference*, pp. 1684–1689.

Feldman, K., Sandborn, P. (2007). Integrating technology obsolescence considerations into product design planning. In *Proceedings of the ASME 2007 International Design Engineering Technical Conferences & Computers and Information in Engineering Conference IDETC/CIE 2007 September 4–7, Las Vegas, Nevada, USA*.

Pantano, E. et al. (2013). Obsolescence risk in advanced technologies for retailing: A management perspective. *Journal of Retailing and Consumer Services, 20*, 225–233.

[7]Robins, N. (2014). Integrating environmental risks into asset valuations: The potential for stranded assets and the implications for long-term investors. International Institute for Sustainable Development.

Uhrynuk, M., & Burdulia, A. W. (2020). Singapore regulator issues environmental risk management guidelines for asset managers and other financial institutions. Mayer Brown 18 Dec 2020.

environmental sustainability requires the integration of environmental risk factors in the funding and financing of investment decisions that culminate in the creation, establishment, utilisation (i.e., operating and maintaining) and retirement of engineered assets.

Notably, environmental impact assessment (EIA) is a practice that has evolved from the environmental sustainability imperative. According to the International Association for Impact Assessment, EIA involves "identifying, predicting, evaluating and mitigating the biophysical, social and other relevant effects of proposed development proposals prior to major decisions being taken and commitments made." Despite the requirement to conduct EIA prior to major decisions being taken and commitments made, however, the pragmatic viewpoint is that EIA involves a systematic process of examining the environmental consequences of acquiring, utilising and retiring engineered assets. Thus, the conduct of EIA should not only occur in the concept phase during asset acquisition, rather EIA should continuously form part of the holistic assessment of the condition of an asset.

6.1.4 Socio-Economic and Socio-Political Considerations

Holistically, the ramifications of the value ethos and the sustainability imperative also imply that socio-economic and socio-political issues are considered during acquisition and throughout the life stages of an asset. Socio-economic factors[8] may be strongly correlated to funding and financing issues, whereas socio-political[9] factors may be more closely related to environmental issues. In conjunction with the environmental impacts, the establishment and deployment of an engineered asset in a particular location imposes consequences on the socio-economic and socio-political arrangements within the surrounding communities.[10] For instance, the establishment of a manufacturing plant influences economic activities and the politics of the community within which the asset is located. It is not unusual for a new community to evolve especially if the manufacturing plant is being established in an area where there is no pre-existing human settlement. In fact, this phenomenon is relatively common in developing countries as policies on sustainable development are in-part informed by socio-economic and socio-political considerations of establishing industrial asset systems.[11]

---

[8]Socio-economic issues are factors that have negative influence on an individuals' economic activity including: lack of education, cultural and religious discrimination, overpopulation, unemployment and corruption.

[9]Socio-political issues range from homelessness, to discrimination, to immigration and the refugee crisis, to health care, and beyond.

[10]Understanding socio-economic and political factors to impact policy change (English). Washington, D.C.: World Bank Group. http://documents.worldbank.org/curated/en/489651468324550090/Understanding-socio-economic-and-political-factors-to-impact-policy-change.

[11]Frederiksen, T. (2019). Political settlements, the mining industry and corporate social responsibility in developing countries. *The Extractive Industries and Society, 6*(1), 162–170.

6.2   Asset Acquisition Options I

This section includes discourse on a number of options for acquiring assets. An asset or a system of assets may be acquired, for instance, through outright ownership, or through hire/lease, and/or partnership arrangements. These options apply to the acquisition of assets in both brown- and green-field environments, albeit that the options are best defined and interpreted in the legal context.

6.2.1   Outright Ownership

Ownership confers title and the widest exclusive rights to an asset; whereas custodianship, or guardianship, or stewardship confers a limited set of rights of control over, as well as accountability and responsibility for an asset. In this sense, employees in a business organisation may be regarded either as custodians, or guardians, or stewards of the company's assets, and their limited rights to the company's assets may be stipulated in both the conditions of employment and job specifications. It is important to note that ownership acquisition of an asset is typically covered by some form of contract agreement which typically states the implied coextensive indemnities. Ownership of an asset can be conferred as a gift (such as bequeathed inheritance), or through direct acquisition by an individual or a legal person such as a business organisation.

Most persons possess assets such as personal gadgets, residential property and associated assets. It is conceivable that socio-economic, funding/financing, technology, environmental, and socio-political issues saliently influenced the decisions for outright ownership of an asset by a person. Curiously, it is not uncommon for individuals to adopt a *laissez faire* instead of a formalised approach to manage personal assets that are not deployed for commercial purposes. Organisations may also have outright ownership of engineered assets and are generally expected to adopt a more formal approach to managing such assets. In principle, the owner of an asset may assign custodianship, guardianship or stewardship rights to individuals and organisations. The terms of assignment may be such that the custodian, guardian or steward has responsibilities for managing the asset. In essence, investors, shareholders and other stakeholders implicitly assign the business organisation the right to manage assets so as to realise value (benefits) to all stakeholders.

6.2.2   Hire/Lease

There are circumstances where it is preferable to acquire an asset through a hire (short term) or lease (long term) arrangement. Notwithstanding the reasons for such preferences, however, the scope of the arrangement is typically stipulated in a contract agreement, and the terms and conditions generally state the coextensive accountabilities, responsibilities and indemnities. This implies that those involved in managing the asset should be conversant with their obligations as stated terms and conditions of the hire or lease, and indeed, in any other terms of agreement stipulated in a contract.

6.3 Asset Acquisition Options II: Partnerships

In the context of this book, a partnership is a formal arrangement by two or more parties to manage an engineered asset. The scope for partnership arrangements is wide ranging and encompasses all the principles and concepts discussed throughout this book. Thus, several types of partnership arrangements may be established between parties interested in ensuring that an asset provides the means for the realisation of value. Literally speaking, the aforementioned options for acquiring assets more or less represent specialised interpretations of partnership arrangements. Irrespective of the connotation, the intrinsic objective of any partnership arrangement is to align the realisation and appropriation of value from an asset. In principle, partnership involves the allocation and sharing of decision making authority, opportunities, resources, and risks. Thus, some of the challenges to partnership arrangements include conflicts of interest, credibility, trust, and value misalignment.

Depending on the legal personality of the parties involved, the option to acquire an asset through a partnership arrangement may be construed either as a (i) public-private partnership (PPP), or (ii) a private-private partnership, eventhough the PPP acronym is commonly understood in terms of (i). Other types of partnership arrangements that involve the acquisition of engineered assets tend to be cloaked in prominent terms like joint-venturing, mergers and acquisitions,[12] outsourcing,[13] and servitisation. An indirect way of acquiring and managing engineered assets is through securitisation.[14] Although securitisation is essentially a financial 'practice of pooling together various types of debt instruments (assets) such as mortgages', collectible art and antiquities, however, the securities are illiquid valuables in the form of engineered assets. For the purposes of this book, the discourse will be limited to public-private partnerships and servitisation.

[12]Malik, F., Anuar, M.A., Khan, S., & Khan, F. (2014). Mergers and acquisitions: A conceptual review. *International Journal of Accounting and Financial Reporting, 1*(1), 520 https://doi.org/10.5296/ijafr.v4i2.6623.

[13]Quinn, J. B., Hilmer, F. G. (1994) Strategic outsourcing. *Sloan Management Review, 35*(4), 43. Cambridge.

[14]Schwarcz, S. L. (1994). The alchemy of asset securitization. *Stanford Journal of Law Business and Finance, 1*, 133.

Vink, D., & Thibault, A. (2007). An empirical analysis of asset-backed securitization. *SSRN Electronic Journal.* https://doi.org/10.2139/ssrn.1014071.

6.3.1   Public-Private Partnerships

The prevalence of public-private partnerships[15] emphasise the participation of private enterprise in the development and operation of public infrastructure. PPPs involve concession agreements that take into consideration the type of infrastructure, financing/investment arrangements and the regimes for operating the resulting assets. The focus of public-private partnerships on infrastructure manifests in several types of PPP concessions, some of which are summarily listed as follows:-

- BOT—public owns; private investor builds, acquires right to operate for a period, then transfers operating right to public; project financing is preferred.
- BOOT—private builds, owns, operates for period, then transfers to public ownership; better financing guarantees.
- BOO—similar to BOOT, except that the useful life = operating period of the asset.
- BLT—build, lease, transfer; a corporation is established to manage the leasing of the publicly owned asset.
- DBFO—similar to BOT; design, build, finance, operate; public manages but concessionaire obtains retribution via toll payments.
- DCMF—design, construct, manage, and finance; similar to DBFO but management is transferred to concessionaire (e.g., prisons and hospitals).

6.3.2   Servitisation

A widely adopted but muted form of partnership arrangement describes the relationship between the asset user/client on the one hand, and the asset manufacturer, supplier, systems integrator, or the vendor on the other hand. The option to acquire an asset through the servitisation option stems from the phenomenon whereby designers, manufacturers, vendors, value-adding resellers, suppliers and agents are increasingly choosing to supplement existing and new products (assets) with service offerings. The service(s) offered may be in the form of tangible or intangible technology. Servitisation requires a two-way strategy and involves the bundling of tangible and intangible services with tangible artefacts (i.e., products or engineered assets). From the supplier/service provider side, servitisation involves the provisioning of a tangible or intangible service component as an embedded added value to the tangible product (asset) supplied to a client. From the client or user side, servitisation involves the acceptance of the tangible or intangible service component provided as an integral part of the tangible artefact or product that is supplied

---

[15]Cui, C., Liu, Y., Hope, A., & Wang, J. (2018). Review of studies on the public–private partnerships (PPP) for infrastructure projects. *International Journal of Project Management, 36*(5), 773–794.

Ma, L., Li, J., Jin, R., Ke, Y. (2019). A holistic review of public-private partnership literature Published between 2008 and 2018. *Advances in Civil Engineering, 7094653,* 18. https://doi.org/10.1155/2019/7094653.

by the manufacturer or vendor. The terms of agreement in a servitisation arrangement is typically stipulated in a contract.

The existence of a servitisation contract implicitly results in the establishment of a product-service-system[16] (PSS) between the asset manufacturer/supplier/vendor and the client/user of the asset. Interestingly, the scope of the PSS may also be configured into a variety of business models that define the relationship between the client/user of asset and the asset manufacturer/supplier/vendor. As illustrated in Fig. 6.2, servitisation involves a wide scope of arrangements from the viewpoints of both the client/user and the manufacturer/supplier/vendor of an asset.

The hierarchical configuration of an asset is very significant in servitisation arrangements. In general, many conventional and pre-existing servitisation arrangements are concentrated around components, spare parts, subassemblies and subsystems. A major motivation for servitisation agreements is that technology developers, integrators and implementers prefer new business models that enhance their revenue generation through sustained commercialisation of intellectual property. It also appears that clients and asset users are increasingly adopting outsourcing arrangements as the preferred methods of dealing with rapid evolutions in technology combined with funding/financing risks, socio-economic/political issues and concerns about the environment. After all, acquiring an asset means that the burden of managing the asset with these risks is tolerable or acceptable. Why not limit risk exposure by only paying for the service obtainable from an asset?

6.3.3   Dematerialisation

Unlike the options discussed in the preceding sections, the option of dematerialisation is an anomaly in terms of asset acquisition. This option demands an answer to the question as to whether the required function can be achieved with a non-material asset. Implicitly, an engineered asset without corporeality or material/physical tangibility should impose the least impact on the environment and ecology. Therefore, the option of dematerialisation is relevant in the context of sustainability, and it is pragmatic to consider dematerialisation during the conceptualisation phase of acquiring the asset.

---

[16]Moro, Cauchick-Miguel & Mendes. (2020). Product-service systems benefits and barriers: an overview of literature review papers. *International Journal of Industrial Engineering and Management, 11*(1).

Xin, Y., Ojanen, V., & Huiskonen, J. (2017). Empirical studies on product-service systems—A systematic literature review. *Procedia CIRP, 64* 399–404.

Cavalier, S. & Pezzotta, G. (2012). Product–Service systems engineering: State of the art and research challenges. *Computers in Industry, 63* 278–288.

Meier, H., Roy, R. & Seliger, G. (2010). Industrial product-service systems—IPS2. *CIRP Annals —Manufacturing Technology, 59* 607–627.

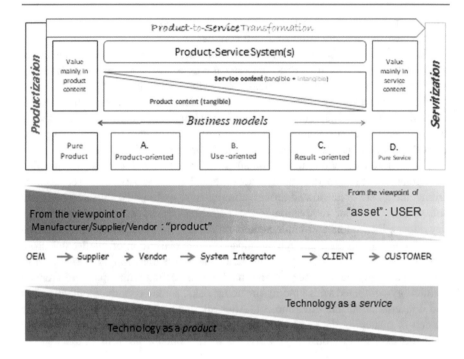

Fig. 6.2 Servitisation and product-service-system

6.4   From Asset Acquisition to Utilisation: Operational Readiness

The funding/financing, technology, environmental, socio-economic and socio-political considerations are not only hugely significant during the acquisition stage but also, their effects tend to be increasingly magnified during the utilisation and retirement stages in the life of an asset. Depending on the size and type of asset, several phases may be required to achieve the acquisition of the asset. Acquiring an asset may only signify that the asset is available for deployment, however, the important question remains as to the extent to which the asset is operationally ready for utilisation.

6.4.1   Asset Acquisition Phases

The resolution of the acquisition stage into a number of phases is conventionally described in the capital project development and implementation plan. An authoritative guideline in this regard is the project management body of knowledge (PMBoK). Depending on the size and type of asset, the resolution may include many phases so as to ensure that asset acquisition is achieved. In the simplest sense, the acquisition of an asset can be resolved into three distinct phases—(i) conceptualisation; (ii) procurement and installation; and (iii) deployment. For brevity, the outcome of the conceptualisation phase may be a bankable feasibility study, whereas deployment results in commissioning the asset as the means to realise particular objectives; e.g., after acquisition, a car may be deployed (i.e. commissioned) either for commercial business, or for pleasure, or for both purposes. Ideally, the cumulative effect of all the activities involved in the acquisition phases should result in an asset that is operationally ready, thus 'operational readiness' marks the transition from acquisition to the utilisation stage in the life of an asset.

6.4.2 Operational Readiness

Irrespective of the size and type of asset, the teaming, resourcing, effort and duration to implement the respective acquisition phases may vary, albeit that particular skills and resources may be sustained through all the phases required to complete the acquisition of the asset. In many asset acquisition situations, the composition of the teams together with the resources employed to achieve the next phase may, at best, have to be reconfigured especially if the skills and competences remain necessary. Otherwise, a new team(s) and resourcing may be established to provide the effort necessary to achieve the next phase(s) in order to complete the process of acquiring the asset.

Incidentally, there is ample empirical evidence in extant literature[17] suggesting that insufficient attention to 'operational readiness' issues, *inter alia*, contribute to the failure of asset acquisition projects either in green-field or brown-field situations. Critically, the handover of the asset needs to be carefully coordinated to ensure that all relevant information (e.g., design, procurement, installation, commissioning, operating, and maintenance details) is passed from the respective team(s) in successive acquisition phases to the asset user. At the end of the acquisition stage, it is vital to ensure that the asset is operationally ready for utilisation. The bar for success is often more rigid in brown-field than in green-field situations. In brown-field situations, the acquired asset must be capable of achieving the pre-established level(s) of utilisation, whereas, in green-field situations, the acquired asset should be capable of achieving the proposed level(s) of utilisation.

Ideally, all operability and maintainability issues (risks) should be explicitly identified and, if possible, resolved as part of ensuring that an asset is operationally ready. At least, the accumulated analyses of hazards, operability and maintainability during the asset acquisition phases should be collated and made available to the team that will manage an asset during the utilisation stage. In principle, operational readiness is about providing seamless transition from the acquisition stage to

---

[17]Guptaa, S.K., Gunasekaranb, A., Antonyc, J., Guptad, S., Bage, S., Roubaudd, D. (2019). Systematic literature review of project failures: Current trends and scope for future research. *Computers and Industrial Engineering, 127,* 274–285.

Hjelmbrekke, H., Hansen, G. K., & Lohne, J.(2015). A motherless child—Why do construction projects fail. *Procedia Economics and Finance, 21,*72–79.

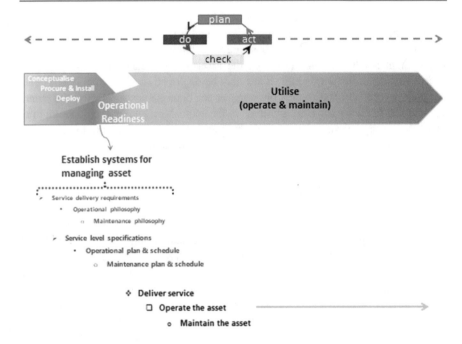

Fig. 6.3 Operational readiness concept

the utilisation stage in the life of an asset. Pragmatically, the transition from acquiring to utilising an asset requires, *inter alia*, the establishment of the asset hierarchy and register as the basis for asset management. As mentioned earlier in Chap. 5, it is absolutely essential to align the engineering and spatial configuration of the asset with the requirements for financial recognition and reporting. At least, operational readiness effort should result in a detailed plan to guide the utilisation of the asset (see Fig. 6.3). The utilisation plan should at least specify the service delivery requirements for the asset in conjunction with the operation and maintenance philosophies. At best, operational readiness should culminate in the integration of the required multidisciplinary skills, competences, together with all the necessary material/non-material resources into a system for managing the asset, i.e., the establishment of a holistic asset management system. In essence, operational readiness is about assuring operational capability of the asset.

6.5 Summary of the Chapter

The main points discussed in this chapter are depicted in Fig. 6.4. The discourse focused on issues that should be considered during the acquisition stage in the life of an asset. Some of the ways of dealing with the risks categorised under funding/financing, technology, environmental and socio-economic/political issues

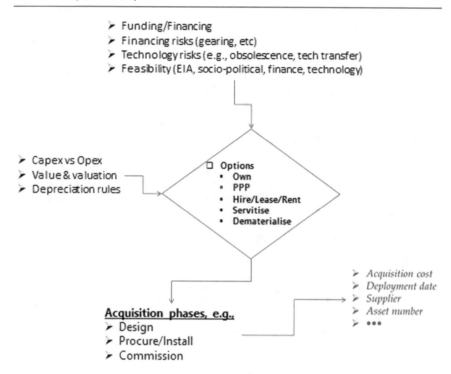

Fig. 6.4 Summary of asset acquisition stage

have been articulated in terms of acquisition options. The strategic significance of various forms of partnerships has been highlighted together with the increasing adoption public-private, servitisation, and outsourcing arrangements. The importance of operational readiness is emphasised as a means of ensuring seamless transition from the acquisition to the utilisation stage in the life of an asset.

6.6   Exercises

Assume that you have been appointed to provide consulting advice to a particular organisation in a particular sector of public or private industry (e.g., agriculture and agro-allied, education and research, health, manufacturing, mining and minerals processing, security, transportation, etc.). Choose the particular organisation and sector.

1. Discuss, in reasonable detail, the challenges that your chosen organisation may confront in acquiring various engineered assets for its business.

2. What options would you recommend for the acquisition of the various engineered assets that the organisation intends to deploy as the means for achieving the business objectives?
3. Explain with conviction why operational readiness is essential for the organisation to transition from acquiring to utilising the assets.

References and Additional Reading

Blanchard, B. S., Verma, D. c., & Peterson, E. L. (1995). Maintainability: A Key to Effective Serviceability and Maintenance Management. Wiley. ISBN: 978-0-471-59132-0.

Cavalier, S. S., & Pezzotta, G. (2012). Product-service systems engineering: State of the art and research challenges. *Computers in Industry, 63*(2012), 278–288.

Cui, C., Liu, Y., Hope, A., & Wang, J. (2018). Review of studies on the public–private partnerships (PPP) for infrastructure projects. *International Journal of Project Management, 36*(5), 773–794.

Dugdale, D., & Abdel-Kader, M. (1999). Funding issues in a major strategic project: A case of investment appraisal. *International Journal of Accounting Education, 8*(1), 31–45. ISSN 0963-9284. https://doi.org/10.1080/096392899331026.

Dunjó, J., Fthenakis, V., Vilchez, J. A., & Arnaldos, J. (2009). Hazard and operability (HAZOP) analysis: A literature review. *Journal of Hazardous Materials, 173*(1–3), 19–32.

Feldman, K., & Sandborn, P. (2007). Integrating technology obsolescence considerations into product design planning. In *Proceedings of the ASME 2007 international design engineering technical conferences and computers and information in engineering conference IDETC/CIE 2007* September 4–7. Las Vegas, Nevada, USA.

Frederiksen, T. (2019). Political settlements, the mining industry and corporate social responsibility in developing countries. *The Extractive Industries and Society, 6*(1), 162–170.

Guptaa, S. K., Gunasekaranb, A., Antonyc, J., Guptad, S., Bage, S., & Roubaudd, D. (2019). Systematic literature review of project failures: Current trends and scope for future research. *Computers and Industrial Engineering, 127*(2019), 274–285.

Hjelmbrekke, H., Hansen, G. K., & Lohne, J. (2015). A motherless child—Why do construction projects fail. *Procedia Economics and Finance, 21*(2015), 72–79.

IATA (2016) Airline disclosure guide: Aircraft acquisition cost and depreciation. International Air Transport Association/KPMG.

Ingo Walter (ed. 2016). *The infrastructure finance challenge.* Open Book Publishers Cambridge: UK. http://dx.doi.org/10.11647/OBP.0106.

Kayser, D. (2013). Recent research in project finance a commented bibliography. *Procedia Computer Science, 17*(2013), 729–736.

Kotek, L., & Tabas, M. (2012). HAZOP study with qualitative risk Analysis for prioritization of corrective and preventive actions. *Procedia Engineering, 42*, 808–815.

Liang, M., Junning, L., Ruoyu, J., Yongjian, K. (2019). A holistic review of public-private partnership literature published between 2008 and 2018. *Advances in Civil Engineering, 7094653*, 18. https://doi.org/10.1155/2019/7094653.

Malik, F., Anuar, M. A., Khan, S, & Khan, F. (2014). Mergers and acquisitions: A conceptual review. *International Journal of Accounting and Financial Reporting, 1*(1), 520. https://doi.org/10.5296/ijafr.v4i2.6623.

Meier, H., Roy, R., & Seliger, G. (2010). Industrial product-service systems—IPS2. *CIRP Annals—Manufacturing Technology, 59*(2010), 607–627.

Moro, S. R., Cauchick-Miguel, P. A., & Mendes, G. H. S. (2020). Product-service systems benefits and barriers: an overview of literature review papers. *International Journal of Industrial Engineering and Management 11*(1).

Müllner, J. (2017). International project finance: Review and implications for international finance and international business. *Manag Rev Q, 67,* 97–133. https://doi.org/10.1007/s11301-017-0125-3.

Pantano, E., Iazzolino, G., & Migliano, G. (2013). Obsolescence risk in advanced technologies for retailing: A management perspective. *Journal of Retailing and Consumer Services, 20*(2013), 225–233.

Quinn, J. B., & Hilmer, F. G. (1994). Strategic outsourcing. *Sloan Management Review. 35*(4), 43, Cambridge.

Rao, V. (2018). An empirical analysis of the factors that influence infrastructure project financing by banks in select asian economies. Asian Development Bank Economics Working Paper No. 554.

Robins, N. (2014). Integrating environmental risks into asset valuations: The potential for stranded assets and the implications for long-term investors. International Institute for Sustainable Development.

Sandborn, P. (2007). In G. Bayraksan, W. Lin, Y. Son, & R. Wysk (Eds.), *Designing for technology obsolescence management proceedings of the 2007 industrial engineering research conference.* pp. 1684–1689.

Schwarcz, S. L. (1994). The alchemy of asset securitization. *Stanford Journal of Law, Business and Finance, 1,* 133.

Uhrynuk, M., & Burdulia, A. W. (2020). Singapore regulator issues environmental risk management guidelines for asset managers and other financial institutions. Mayer Brown 18 Dec 2020.

Vink, D., & Thibeault, A. (2007). An empirical analysis of asset-backed securitization. *SSRN Electronic Journal.l* https://doi.org/10.2139/ssrn.1014071.

World Bank Group (2013) *Understanding socio-economic and political factors to impact policy change* Washington:D.C. http://documents.worldbank.org/curated/en/489651468324550090/Understanding-socio-economic-and-political-factors-to-impact-policy-change.

Xi, G. (2017). The impact of maintainability on the manufacturing system architecture. *International Journal of Production Research, 55*(15), 4392–4410, https://doi.org/10.1080/00207543.2016.1254356.

Xin, Y., Ojanen, V., & Huiskonen, J. (2017). Empirical studies on product-service systems—A systematic literature review. *Procedia CIRP, 64*(2017), 399–404.

Yussof, S. H., Isa, K., & Mohdali, R. (2014). An analysis of the gap between accounting depreciation and tax capital allowance in Malaysia. *Procedia–Social and Behavioral Sciences, 164,* 351–357.

Zaidi, S., Kaz,i A.M., Riaz, A., Ali, A., Najmi, R., Jabeen, R., Khudadad, U., & Sayani, S. (2020). Operability, usefulness, and task-technology fit of an mHealth app for delivering primary health care services by community health workers in underserved areas of Pakistan and Afghanistan: Qualitative study. *Journal of Medical Internet Research, 22*(9), e18414, https://doi.org/10.2196/18414

Zhu, L., Shan, M., & Bon-Gang Hwang, M.ASCE, (2018). Overview of design for maintainability in building and construction research. *Journal of Performance of Constructed Facilities, 32*(1), (February 2018)

Asset Utilisation Stage

7

Abstract

The discourse in this chapter is about issues concerning the utilisation stage in the life of an engineered asset. As indicated in the previous chapter, funding and financing, technology, environment and socio-economics/politics issues also prevail during the utilisation stage. The focus here is on issues regarding operations, maintenance, condition and performance assessments, as well as asset management decisions arising thereof.

7.1 Operational Readiness to Utilisation

The purpose of operational readiness is to ensure seamless transition from the acquisition to the utilisation stage in the life of an asset. In this regard, the extent to which the system(s) for managing an asset have been established is tantamount to the level of readiness to utilise an asset, i.e., to operate, maintain, and keep track of the condition and performance of an asset. That is, operational readiness establishes an overarching system for managing an asset. Thus the management system encompasses, *inter alia*, the extent to which the

(i)   necessary multidisciplinary skills and competences have been integrated with all the required material/non-material resources;

(ii) engineering and/or spatial configuration of the asset has been aligned with the requirements for financial recognition and reporting; and

(iii) service delivery requirements, plus the operational and maintenance philosophies have been translated into a comprehensive utilisation plan, i.e., workable plans and schedules of activities that will be carried out as the asset is being utilised.

7.1.1 Operating Philosophy

In practice, it is essential that the service delivery requirements correspond to the actual, rather than some perceived deployment of an asset. For instance, a motorcar can be deployed either as a taxi or for any other purpose, so the service delivery requirements must be stipulated according to the purpose. To determine the utilisation plan and schedule, it is vital to translate the service delivery requirements (i.e., the time-dependent profile of demand for an asset) into an operating philosophy that guides how an asset will be operated and maintained, including choices that will be made as the asset is being utilized.

7.1.2 Operational Planning and Scheduling

Ideally, the operating philosophy should ensure that the asset is capable of delivering the required levels of functional performance while the asset is in use. The operating philosophy must be translated into operations plans and schedules that match the time-dependent demand profile for the asset against measurable parameters such as availability, capacity, energy consumption and throughput. Operational planning and scheduling[1] take into consideration technical factors such as the demand loading and duty cycle, the range of permissible operating regimes plus either the time- availability or capacity-availability of the asset or both, as well as reliability, integrity and energy consumption of the asset.

Other factors to be considered include scheduled shutdown periods that allow for planned and scheduled maintenance activities to be carried out plus specific procedures for dealing with abnormal operating conditions, unscheduled shutdowns, and return-to-service challenges. Non-technical factors include, *inter alia*, operational staffing and resource requirements, and costs. In practice, operations plans and schedules are typically stipulated as standard operating procedures (SOPs), that is, step-by-step instructions designed to enable operators 'to achieve efficiency, quality output and uniformity of performance, while reducing miscommunication and failure to comply with industry regulations.'

Detailed operations planning and scheduling depends on the asset and the purpose(s) of the asset. That is, the level of detail and formality of an operations plan and schedule will vary depending on the complexities associated with the asset and its purpose(s). Readers interested in treatise on detailed operations planning and scheduling may refer to extant literature regarding specific assets in specific industrial sectors.[2]

[1]Ahmed, M. G. (2020). An optimization model for operational planning and turnaround maintenance scheduling of oil and gas supply chain. *Applications Science, 10*, 7531. https://doi. org/10.3390/app10217531.

[2]Vuchic, V. R. (2005). Urban transit: Operations, planning and economics. Wiley. ISBN 0– 471-63265-1.

Bouffard, F., & Galiana, D. G. (2008). Stochastic security for operations planning with significant wind power generation. *IEEE Transactions on Power Systems, 23*(2), 306-361.

7.2 Maintenance

A puzzling culture prevails in practice in the sense that operating and maintaining an asset are compartmentalised and regarded in organisational structure 'silos' as if they are mutually exclusive functions. The reality is that operating and maintaining an asset are two sides of the same coin, that is, operators and maintainers are twins. It is remarkable that the fanciful term 'asset care' reflects the fact that operating an asset is twined with maintaining the asset. The notion of 'asset care' connotes that the maintenance philosophy derives from the operations philosophy and vice versa. After all, if an asset is not operated, it may be difficult to justify doing particular maintenance tasks. Thus, it is pragmatic that maintenance planning and scheduling is subsequent to operations planning and scheduling, even though both processes of operations and manitenance planning and scheduling can be iteratively push-pull.

7.2.1 Maintenance Philosophy

The philosophy of maintenance is rooted in the risk of technical malfunction or functional failure of an asset. The primary question is whether or not the risk of technical malfunction or functional failure of an asset is *acceptable* or *tolerable*. Maintenance philosophy has been translated into the normalised[3] conventions depicted in Fig. 7.1. Although *adhoc* maintenance tends to be muted, however, it represents the most instinctive approach especially in situations where the risk of technical malfunction or functional failure of an asset may be tolerated or even accepted.

7.2.2 Maintenance Conventions

Reactive maintenance more or less implies that the risk of technical malfunction or functional failure may be *tolerable* but, not necessarily acceptable. Preventative, predictive, and proactive maintenance conventions are founded on the basis that the risk of technical malfunction or functional failure of an asset is unacceptable!

The absolute preventative maintenance convention presumes that, under specified operating conditions and regimes, the occurrence of technical malfunction

Mateljak, Z., & Mihanović, D. (2016) Operational planning level of development in production enterprises in the machine building industry and its impact on the effectiveness of production. *Economic Research-Ekonomska Istraživanja, 29*(1), 325–342, https://doi.org/10.1080/1331677x.2016.1168041.
[3]CEN—EN 13306 Maintenance—Maintenance terminology
The Maintenance Framework. Global Forum on Maintenance and Asset Management (GFMAM) (2016). ISBN: 978–0-9870602-5-9.

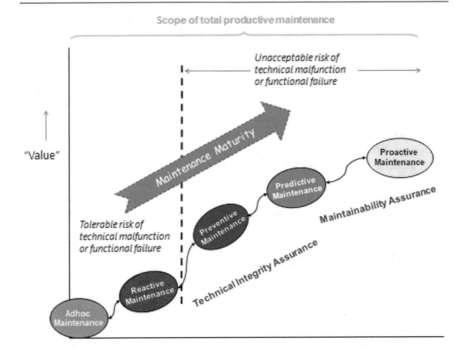

Fig. 7.1   Maintenance philosophy and conventions

or the failure modes[4] inherent in an asset, as well as the times when the respective malfunctions or failures will occur, are known. Hence, definitive preventative maintenance tasks may be specified *apriori* or even *ab initio*. Predictive maintenance assumes that, under specified operating conditions and regimes, the occurrence of technical malfunction or the failure modes inherent in an asset, as well as the times when the respective malfunctions or failures may occur, can be estimated or reasonably ascertained by monitoring (i.e., measuring and tracking) and analysing signals that can reveal the incidence or latency of technical malfunctions or failure modes.

In situations where the operating conditions are presumed to be ergodic and statistically stationary, it is possible to know beforehand that a particular malfunction and/or mode(s) of failure will occur but, the time of occurrence may not be known. The monitoring and analysis of pertinent signals may be focused on ascertaining the time when the malfunction/failure mode(s) will most likely occur. This is tantamount to measuring the reliability of an asset. In situations where the operating conditions are either generally not ergodic/non-stationary, or there is anticipation that supervening events may occur during a particular time period, the

---

[4]BS EN IEC 60812:2018 Failure modes and effects analysis (FMEA and FMECA).
  Cristea, G., & Constantinescu, D. M. (2017). In IOP Conference Series: Materials Science and Engineering. Vol 252. pp. 012046.

monitoring and analysis of pertaining signals may be focused on the prognosis of the malfunction/modes of failure that will most likely occur at particular times. This extends the scope of reliability measurement to assessment of the technical integrity and robustness of an asset.

Preventative and predictive maintenance not only mark a significant shift in risk intolerance but also represents a focus towards assuring the reliability of an asset under stationary operating conditions, and the resilience and technical integrity of an asset under non-stationary operating regimes. In the same vein, preventative and predictive maintenance conventions point towards maintainability assurance during the utilisation stage in the life of an asset. Whereas preventative and predictive conventions are about coping with uncertainty and risk ambiguity associated with malfunctions and modes of failure of an asset, however, proactive maintenance is about eliminating troublesome modes of failure, i.e., modes of failure that occur more frequently with increasing severity under normal (i.e., ergodic and stationary) operating conditions. In reality, the complete elimination of a particular mode of failure may only be achieved by redesigning components, spare parts, subassemblies, subsystems, or even the asset as may be deemed necessary.

Maintenance maturity generally refers to the transition from *adhoc* and reactive conventions to preventative, predictive and proactive approaches. In maintenance practice, the maturity process involves auditing and benchmarking to establish the prevailing conventions, then to design a roadmap for transitioning to higher value conventions. Suppose that an audit and benchmark exercise reveals that *adhoc* and reactive conventions are prevalent, this situation may trigger the initiation of projects that involve design and implementation of a roadmap to transition to the more risk intolerant conventions. On the one hand, technology advancements are resulting in digitalised twins of tangible assets. On the other hand, the tangible components, subassemblies, subsystems and assets are becoming highly reliable because, not only are they being made with innovative materials but also, they are being embedded with automated self-checking and IoT capabilities. Hence, maintenance maturity may be more about mixing the reactive, preventative, and predictive conventions in order to achieve the desired levels of improvement in overall reliability, robustness and technical integrity of an asset.

7.2.3 Maintenance Planning, Scheduling, and Execution

As a matter of course, maintenance plans and schedules derive from the operations plans and schedules. That is, the plan and schedule for maintaining an asset should actually be done only after the operation plan and schedule for a specified time period must have been established. After all, any changes in the operation plan and schedule will invariably necessitate corresponding changes in the maintenance plan and schedule. To create a maintenance plan and schedule, the reader may refer to the earlier discourse on practical concepts in Chap. 4.

The maintenance of an asset can be achieved through intervention, investigative or restorative actions. In some instances, simple check or inspect actions can be carried out with the asset in operation, otherwise most investigative, intervention and/or restorative actions require that the asset is in a power-down or shutdown mode. Infact, all maintenance actions are corrective in the sense that an intervention task can be done to prevent the occurence of a deviation, or a restoration task can be done after a deviation has occurred. In effect, preventative and predictive maintenance actions incorporate investigative actions which involve monitoring signals that provide some indication as to the operational environment surrounding an asset.

7.3 Condition Monitoring, Diagnostics and Prognostics

The processes of technical integrity and maintainability assurance depend on measuring and tracking signals that can reveal the incidence of malfunction and failure modes inherent in an asset. During the acquisition stage (especially during the installation and commissioning phases), such signals can be measured and tracked whilst an asset is subjected to test conditions, e.g., for factory or site acceptance, and for purposes of operational readiness. The installation and commissioning test situations (including non-destructive testing) are often used to establish a baseline for subsequent monitoring of the failure characteristics especially during the utilisation stage of an asset. Non-destructive testing (NDT) has long been established as a method for detecting anomalies and defects in engineered assets. Although traditional NDT techniques such as acoustic emission and eddy current testing, magnetic particle inspection, radiographic and ultrasonic testing are typically applied to detect manufacturing anomalies and defects, however, the techniques are increasingly being applied for condition monitoring throughout the life stages of engineered assets.

Remarkably, there may be situations where the technical malfunction of a component, spare part, subassembly, or subsystem may be tolerated before such malfunction induces a functional failure of the asset. Conventional or technical condition monitoring[5] generally refers to the measurement and tracking of signals that can reveal the incidence or latency of malfunction or the failure modes of an asset. Typically, the monitoring process involves taking measurements of signals from devices that sense variables such as pressure, temperature, and flow, as well as physical parameters like acidity/alkalinity (pH), dissolved oxygen, thermo-electric effect, sound, vibration, etc. This is followed by analyses of the measured sensor

---

[5]ISO 17359, Condition monitoring and diagnostics of machines—General guidelines.

ISO 13373, Mechanical vibration and shock—Vibration condition monitoring of machines.

ISO 13379, Data interpretation and diagnostic techniques which use information and data related to the condition of a machine.

ISO 13381, Condition monitoring and diagnostics of machines—Prognostics.

ISO 18436 Part II, Accreditation of organizations and training and certification of personnel—Part II—General requirements for training and certification—Vibration Analysis.

ISO 18436 Sub Parts under development for Oil Analysis and Thermal Imaging.

signals to identify changes that may reveal the incidence, latency or occurrence of technical malfunction or functional failure. Furthermore, changes in the sensor signals may be analyzed so as to predict the likelihood of technical malfunction or functional failure an asset.

The analyses methods generally depend on the nature of the signals produced by the sensors, especially on the statistical characteristics of the signals. With advances in materials technology, it is now possible, not only to embed sensors internally within components, spare parts, subassemblies and subsystems but also, to install many sensors at various levels of an asset hierarchy and to obtain large sets of data from the sensor signals. The collection of data from a variety of sensors and other sources, in large volumes and at very high rates, results in big data which generally require complex, sophisticated and specific analytical techniques in order to extract useful and valuable information. Notably, artificial intelligence algorithms are being increasingly applied to extract 'knowledge' from such big data.

Where changes in the sensor signals indicate that technical malfunction or functional failure has occurred, then further investigative maintenance actions can be initiated in order to diagnose the cause of failure. This is tantamount to identifying the hazards or sources of risk of the malfunction/failure incident. Furthermore, changes in the sensor signals can be projected to forecast the likelihood of defect occurence. Where changes in the signals reveal that technical malfunction or functional failure may be imminent, such information can then be extrapolated to indicate when technical malfunction or functional failure will likely occur (this is tantamount to estimating or predicting the end-of-life of an asset due functional failure). Prognostic maintenance actions may then be initiated to avert, or if possible, eliminate the likelihood of failure within the forecast time period.

In essence, data and information obtained from monitoring the sensor signals may trigger diagnostic, concurrent intervention and/or restorative or investigative prognostic maintenance actions. Furthermore, the condition monitoring data and information necessitate updates to maintenance plans and schedules, as encapsulated within the prevailing operations plans and schedules, and may trigger updates to operations philosophy to keep track with intransigent and evolving conditions surrounding the asset. Maintenance actions arising from preventative, predictive and prognostic conventions are often regarded as condition-based maintenance. This is suboptimal in the sense that circumstances which cause the maintenance plans and schedules to be misaligned to the operations plans and schedules deserve much more holistic assessment of the condition of the asset.

7.4 Condition and Performance Assessments

The paramount concern during the utilisation stage is that an asset must be in a condition to guarantee continuous realisation of quantitative and qualitative aspects of value. As discussed earlier in §2.2, there are various stakeholders to an asset and each stakeholder has a biased perception of the quantitative and qualitative aspects of the value that should be realised from an asset. For instance, investors transform to shareholders and value projections must, at least, be translated from investment appraisal projections into return on investment dividends during the asset utilisation stage. Hence, economic profit is typically the dominant value demanded by the shareholders[6]. Similarly, other stakeholders expect, and sometimes demand that their value projections are realised during the utilisation of stage an asset. The reporting requirements are typically stipulated in the various standards discussed earlier in Chap. 5.

7.4.1 Asset Condition and Value

Irrespective of the value desired by the various stakeholders, the reality is that utilising an asset also causes the asset's technical condition to deteriorate[7], concomitant with degradation in performance. Thus, the realisation of future value depends on the current overall condition of the asset. In general, the deterioration of an asset could be due to ageing mechanisms plus utilisation patterns, as well as the interaction between ageing and utilisation. For this reason, it is pertinent to holistically determine the condition of an asset at a particular point in time to inform future decision-making. Such holistic determination of asset condition extends beyond adjusting maintenance tactics and can be elevated to changes in utilisation patterns as follows.

First, depending on the configuration, even if monitoring of corresponding sensor signals indicates deterioration in some components, spare parts, subassemblies or subsystems, yet the overall functioning of an asset may be such that the asset still performs at an acceptable level, especially where the level of performance is acknowledged by stakeholder whose value propositions are paramount at the time of assessment. Second, the monitoring of signals to ensure that maintenance actions are aligned to the prevailing operating regimes can occur on a continuous basis, whereas condition assessment is a snapshot in time determination as to whether an asset can deliver desired levels of performance. Third, knowledge accumulated by trending the monitored signals can trigger or inform the need to conduct an encompassing assessment of an asset's condition at a particular point in time.

The overriding assessment question is whether an asset is in a condition to provide the means for the realisation of future value. Given the particular point in time, is an asset in a condition to deliver future levels of service demanded in a manner that achieves the so-called balancing of "costs, risks and performance[8]"?

---

[6]Shareholders represent only one group of stakeholders.

[7]Sanchez-Silva, M., Klutke, G.-A., & Rosowsky, D. V. (2011). Life-cycle performance of structures subject to multiple deterioration mechanisms. *Structural Safety, 33*(3), 206–217. ISSN 0167-4730. https://doi.org/10.1016/j.strusafe.2011.03.003.

Renwick, D. V., & Rotert, K. (2019). Deteriorating water distribution systems can impact public health. *Opflow, 45*(9), 12–15. https://doi.org/10.1002/opfl.1246.

[8]Re: ISO 5500 × definition of an asset.

The answer to this question can be obtained by conducting a holistic assessment of the asset condition in terms of the value ethos and the sustainability imperative.

A holistic assessment involves environmental, financial, functional, socio-political and technical examination in order to establish the overall condition of an asset. It is remarkable that engineering/technology stakeholders tend to emphasise the technical dimension, similarly, other disciplines tend to emphasise the particular dimension of condition assessment most aligned to their biases and persuasions. Hence, to operationalize the value ethos and the sustainability imperative, five dimensions to condition assessment are proposed in this book.

7.4.2 Condition Assessment Dimensions

Data, information and knowledge regarding the value ethos and the sustainability imperative, together with the principles of risk, resilience, sustainability and vulnerability may be collated into five non-exclusive dimensions in order to provide a holistic description of the condition of an asset. The five dimensions for condition assessment are illustrated in Fig. 7.2. The extent to which the dimensions are correlated depends on whether the corresponding factors and parameters are latent, symptomatic, or primary indicators of asset condition. Obviously, the factors tend to be qualitative while the parameters are quantitative.

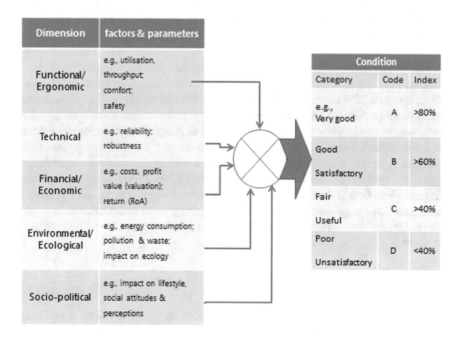

Fig. 7.2 Five dimensions for holistic assessment of asset condition

- The functional/ergonomic dimension provides a snapshot indication of how the asset functionally performs the demanded or specified level(s) of service at a given point in time. The relevant indicators[9] of functional and ergonomic condition of an asset include, for example, qualitative factors like levels of comfort and safety, and a quantitative parameter like throughput.
- The technical dimension provides a snapshot indication of the stress handling capability of an asset. This dimension takes into consideration that technical stressors can arise from endogenous and exogenous sources of risks and uncertainty. Parameters like reliability and robustness (resilience) representing this dimension are generally quantitative. In the evolving parlance of engineering asset management, the technical dimension is often regarded as asset health monitoring, meaning that a 'healthy' asset should be in a condition to perform as expected.
- The economic/financial dimension is relatively easy to assess. Cost, profit and cost-benefit ratios are quantities that can be readily assessed at a snapshot in time. These verifiable quantities are usually obtained *post facto*, so, it is not uncommon that decision-making tends to be biased towards economic and financial indicators of asset condition.
- The environmental/ecological dimension has taken on heightened significance as a consequence of the sustainability imperative. Energy consumption, material content, pollution, waste, and the way in which an asset impacts on ecological biochemistry and cycles have become issues of great concern for asset management.
- Pollution and waste can also be linked to socio-political assessment factors like social attitudes towards an asset, perceptions of an asset by the surrounding community and wider society, as well as the real impacts of an asset on the lifestyles of the surrounding community and wider society.

It is worth reiterating that some of the factors and parameters mentioned above may be resolved into various types of data and information that can be obtained about an asset. The scope of data and information that can be collected for monitoring and assessment of asset condition range from sensor signals of physical variables, to spatial variables typically captured in geographic information systems, to information obtained from visual observations, and empirical responses from interviews, survey questionnaires, *et cetera*. It is pragmatic to convert information about the qualitative factors into quantities (e.g., via Likert scale surveys into descriptive statistical data) so as to assign numerical weights to all the factors and parameters. As suggested in Fig. 7.2, the weighting of the respective factors and parameters can be aggregated to provide a numerical indication of the holistic asset condition.

---

[9]Neumann, P., Kihlberg, S., Medbo, P., Mathiassen, S. E., & Winkel, J. (2002). A case study evaluating the ergonomic and productivity impacts of partial automation strategies in the electronics industry. *International Journal of Production Research, 40*(16), 4059-4075. https://doi. org/10.1080/00207540210148862.

There are many methods for conducting and reporting the condition of various categories and types of assets. As discussed in extant literature[10], there are also numerous case studies on the technical dimension of condition assessment for different categories and types of assets. Each case study tends to focus on a particular type of asset and how the asset is utilised. The monitoring of signals and the assessment of asset condition depend on the disciplines involved, the industry sector concerned, and the applicable standards. 4IR technologies are increasingly applied to obtain various types of data and information for monitoring and assessment of asset condition. Although there are many purported claims of applications in marketing brochures, nevertheless, there are also some substantive case studies[11] of applications of 4IR technology towards monitoring and assessment of asset condition.

7.4.3   Performance Assessment

In acknowledging that the condition of an asset determines value that can be obtained during the utilisation stage, it therefore implies that a holistic assessment of an asset's condition is tantamount to measuring the performance of managing the asset. On the

---

[10]Guidelines for condition assessment of educational facilities. Version 1, 2019. Department of Basic Education, Republic of South Africa.

Paik, J. K., Melchers, R. E. (eds). (2008). In Condition assessment of aged structures. CRC Press.

Humberto, E. G., Wen-Chiao, L., & Semyon, M. M. (2012). A resilient condition assessment monitoring system. https://core.ac.uk/download/pdf/190695757.pdf.

David, M., & Burn, S. (2008). Effective use of condition assessment within asset management. *Journal—American Water Works Association, 100*(1), 54-63. https://doi.org/10.1002/j.1551-8833. 2008.tb08129.x.

Mayo, G., & Karanja, P. (2018). Building condition assessments—Methods and metrics. *Journal of Facility Management Education and Research, 2* (1), 1–11. https://doi.org/10.22361/jfmer/91666.

Wahida, R. N., Milton, G., Hamadan, N., Lah, N. M. I. B. N., Mohammed, M. A. H. B. (2012). Building condition assessment imperative and process. *Procedia—Social and Behavioral Sciences, 65*, 775–780 https://doi.org/10.1016/j.sbspro.2012.11.198.

Marcia, L. B., Ivo, P. de M., Joao, P. M., Cairo, L. N., Helmut, P., Elsa, H. (2017). Aircraft on-condition reliability assessment based on data-intensive analytics. In 2017 IEEE Aerospace Conference. IEEE Xplore https://doi.org/10.1109/AERO.2017.7943870.

ICAO - The New Global Reporting Format for Runway Surface Conditions.

Syaza I., Ahmada, B., Haslenda Hashima, B., Hassim, M. H. (2014). Numerical descriptive inherent safety technique (NuDIST) for inherent safety assessment in petrochemical industry. *Process Safety and Environmental Protection, 9*(2), 379–389.

Pellegrino, J. L., & Carole, T. M. (2004). Impacts of condition assessment on energy use: Selected applications in chemical processing and petroleum refining. Report prepared for U.S. Department of Energy Industrial Technologies Program Washington, D.C

Parida, A. (2011). Condition monitoring and maintenance performance assessment issues for mining industry. *International Journal of COMADEM, 14*(2), 27–34.

Parida, A., & Phillip, T. (2017). Condition monitoring and diagnosis of modern dynamic complex systems using criticality aspect of key performance indicators. *International Journal of COMADEM, 20*(1), 35–39.

[11]Stankovic, M., Hasanbeigi, A., & Neftenov, N. (2020). In Marcello, B., Anamaría, N., Raphaëlle, O. (Eds.), Use of 4IR technologies in water and sanitation in Latin America and the Caribbean. IDB-TN-1910 Inter-American Development Bank.

one hand, a holistic condition assessment provides a present time snapshot of the capability of an asset but, with a futuristic view. On the other hand, performance assessment is retrospective; it involves a retroactive aggregation of the asset condition over a historical period of time. Hence, performance assessment is about aggregating the results of more than one instance of holistic condition assessment.

It is useful to distinguish *asset performance* from *asset management performance*. Asset performance can be established under strictly controlled conditions such as during experimentation, testing or simulation. Thus, asset performance is closely related to the technical dimension of condition assessment, assuming that stressors attributable to the other dimensions remain stationary. After all, the expectation is that the functional performance of an asset under controlled test or simulated conditions should replicate during actual use conditions. The reality can be quite different as experimental, test or simulated conditions do not always replicate in the real world!

Asset management performance is holistic and includes all elements of the aforementioned five dimensions. Performance assessment is commonly proffered to justify the implementation of business intelligence systems[12]. The picture in Fig. 7.3 depicts a simplified input/out systems model that can be applied for assessing asset management performance. The model essentially expands the ratio of value obtained to value invested into measurable parameters (metrics) attributable to the five dimensions that exert endogenous and exogenous influences on the management of an asset.

Metrics such as return-on-assets and asset effectiveness are regarded as strategic; asset utilisation and throughput are regarded as tactical; whereas reliability is regarded an operational metric. In situations where an asset (e.g., an item of equipment or a machine) or where a system of assets (e.g., a manufacturing or processing facility) is deployed to convert quantifiable inputs into quantifiable outputs, asset effectiveness is often calculated in terms of the following parameters as

$$effectiveness = \prod efficiency \bullet utilisation \bullet yeild \qquad (7.1)$$

The effectiveness metric depends on how an asset hierarchy is defined or configured. Measurements of efficiency and yield may not be trivial even though yield can be readily ascertained in terms of throughput. Instead of availability, utilisation is a more accurate measure in the sense that, if an asset deployed for manufacturing or production is not utilised, such an asset cannot create value.

Another relatively simple but very insightful measure of performance is the economic value added (EVA) by an asset. EVA is generally defined in terms of the

---

[12]Business intelligence systems provide historical, current, and predictive views of business operations, most often using data that has been gathered into a data warehouse or a data mart and occasionally working from operational data. The software includes a set of tools that facilitate retrieval, analyses, and transformation of data and information into useful business insights.

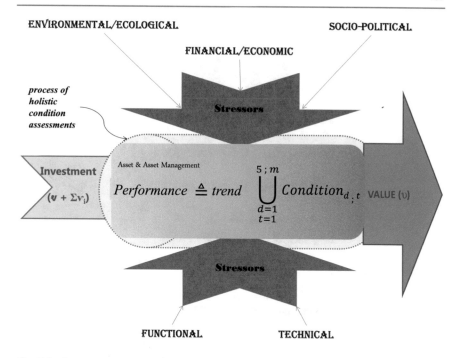

Fig. 7.3 Asset management performance assessments

net investment, actual return on investment, and cost of capital, and can be calculated as per Eq. (7.2) (see also the illustration in Fig. 7.4.)

$$EVA = NOPAT - (WACC * capital invested) \qquad (7.2)$$

Although effectiveness and EVA are highlighted here, however, other metrics of asset management performance can be obtained by aggregating the factors and parameters obtained from condition assessments. It is important that the hierarchy of metrics aligns with the asset hierarchy and conforms to stipulated formats for performance reporting purposes.

7.4.4 Performance Reporting

The requirements for reporting asset condition and asset management performance are usually specified in most of the suites of standards discussed earlier in Chap. 5. It is essential to follow the relevant guidelines and procedures and to comply with pertaining regulations. In practice, poor quality of data and information often bedevils the reporting of asset condition and asset management performance. For instance, inaccuracies in the collection of data (e.g., incorrect sampling of signals

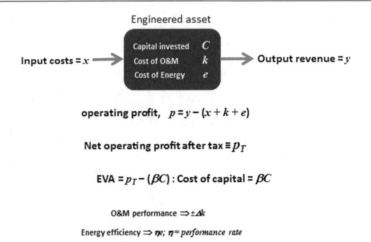

Fig. 7.4 An illustration of EVA calculation

representing condition parameters) plus invalid empirical evidence regarding the factors that influence performance are very significant practical challenges.

Ultimately asset condition and asset management performance assessments constitute the building blocks of the global standards for reporting on economic, environmental and social aspects of a business. It is remarkable that asset management is at the core of the circular economy[13] goals of minimising depletion of natural resources and pollution, and eliminating waste by ensuring the continual use of resources through dematerialisation, asset reuse, recycling and life-extension practices. The global reporting initiative (GRI) emphasises the so-called triple bottom line[14] or triple-helix characterisation of the interactions between people, prosperity and the planet in terms of the sustainability imperative. It is worth commenting that global reporting is sometimes treated as a governance issue possibly because "...the GRI has successfully become institutionalised as the preeminent global framework for voluntary corporate environmental and social reporting[15]." The requirements for reporting in conformance to GRI standards trickle top-down through the various levels of asset management decision making.

---

[13]https://www.europarl.europa.eu/news/en/headlines/economy/20151201STO05603/circular-economy-definition-importance-and-benefits..

Kirchherr, J., Reike, D., & Hekkert, M. (2017). Conceptualizing the circular economy: An analysis of 114 definitions. *Resources, Conservation and Recycling, 127*, 221–232. ISSN 0921-3449. https://doi.org/10.1016/j.resconrec.2017.09.005.

[14]Slaper, T. F., & Hall, T. J. (2011). The triple bottom line: What is it and how does it work? *Indian a Business Review, 86*(1).

Hammer, J., & Pivo, G. (2016). The triple bottom line and sustainable economic development theory and practice. *Economic Development Quarterly*. https://doi.org/10.1177/0891242416674808.

[15]Levy, D., Brown, H. S., & de Jong, M. (2010). The contested politics of corporate governance: The case of the global reporting initiative. Management and Marketing Faculty Publication Series. Paper 1. http://scholarworks.umb.edu/management_marketing_faculty_pubs/1.

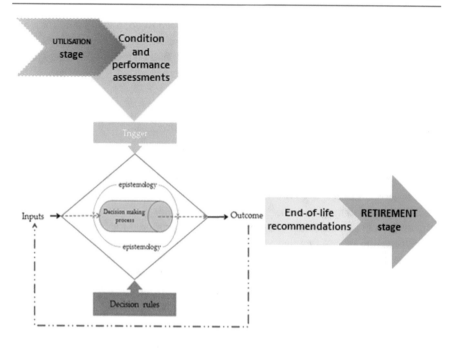

Fig. 7.5 Asset management decisions

7.5   Asset Management Decisions

Ideally, continuous conditioning monitoring should also inform continuous decision making through all the life stages of an asset. However, decisions that result in the transition between the asset life stages are generally subsequent to holistic condition and performance assessments. This is especially the case as the asset transitions between the life stages of utilisation and retirement[16]. As illustrated in Fig. 7.5, condition and performance assessments trigger the decision making process and the outcome is usually 'end-of-life' recommendations which have to be implemented during the retirement stage. Hence, the recommendations regarding end-of-life options more or less mark beginning of the retirement stage in the life of an asset.

The scope of the condition and performance assessments depends on what is defined as the asset in relation to the overall asset hierarchy. Whereas it may be a trivial recommendation to replace a single hardware item like a valve that has failed, however, the recommendation to replace an asset system (e.g., a turbo-machine) will most likely be justified by evidence provided through condition and

---

[16]Dakada, M. H. C., & Amadi-Echendu, J. (2020). Asset replacement decisions in the mining sector. In Liyanage et al (Eds.), Engineering assets and public infrastructure in the age of digitalisation. Springer. ISBN 978-3-030-48020-2. https://doi.org/10.1007/978-3-030-48021-9.

performance assessments. Despite the logical approach of conducting condition and performance assessments to inform asset end-of-life recommendations, it is intriguing that the chosen option may eventually depend on the epistemology of the decision making process.

7.6   Summary of the Chapter

The discourse in this chapter highlighted several issues that are significant during the utilisation stage in the life of an asset. The importance of operational readiness as an orderly transition from the acquisition stage was reiterated. The requirement to resolve the operating and maintenance philosophies into workable plans and schedules were emphasised. Five dimensions were proffered for holistic assessment of asset condition. The fact that asset condition and asset management performance assessments lead to informed decision making was discussed in reasonable detail. Also, the discourse briefly highlighted the fact that asset management is central towards the achievement of circular economy goals as reflected in the global reporting standards.

7.7   Exercises

1. Does operating an asset exclude from maintaining the asset? Explain your answer with succinct arguments.
2. Choose an asset that has been deployed either for a purely commercial profit purpose, or purely for the purpose of non-profit service delivery. Your chosen asset should have some history of being utilised over a minimum period of 5 years. Collect data and information regarding the asset, then

   a. deduce the philosophy that has been applied to operate the asset;
   b. also deduce the maintenance tactics that has been applied to the asset;
   c. determine the trends in the respective dimensions of asset condition assessment;
   d. what would you recommend following the results from (a), (b) and (c)?

References and Additional Reading

Aditya, P. (2011). Condition monitoring and maintenance performance assessment issues for mining industry. *International Journal of COMADEM, 14*(2), 27–34.
Aditya, P., & Tretten, P. (2017). Condition monitoring and diagnosis of modern dynamic complex systems using criticality aspect of key performance indicators. *International Journal of COMADEM, 20*(1), 35–39.

Ahmad, S. I., Hashima, H., & Hassim, M. H. (2014). Numerical descriptive inherent safety technique (NuDIST) for inherent safety assessment in petrochemical industry. *Process Safety and Environmental Protection, 9*(2), 379–389.

Ahmed, M. G. (2020). An optimization model for operational planning and turnaround maintenance scheduling of oil and gas supply chain. *Applications Science, 10*, 7531. https://doi.org/10.3390/app10217531

Bouffard, F., & Galiana, D. G. (2008). Stochastic security for operations planning with significant wind power generation. *IEEE Transactions on Power Systems, 23*(2), 306–361.

BS EN IEC 60812:2018 Failure modes and effects analysis (FMEA and FMECA).

Cawley, P. (2001). Non-destructive testing—current capabilities and future directions. *Part L: Journal of Materials: Design and Applications, 215*(4), 213–223. https://doi.org/10.1177/146442070121500403.

CEN—EN 13306 Maintenance—Maintenance terminology.

Cristea, G., & Constantinescu, D. M. (2017). In IOP conference series: Materials science and engineering. Vol 252. pp. 012046.

Dakada, M. H. C., & Amadi-Echendu, J. (2020). Asset replacement decisions in the mining sector. In Liyanage et al (Eds.), Engineering assets and public infrastructure in the age of digitalisation. Springer. ISBN 978-3-030-48020-2. https://doi.org/10.1007/978-3-030-48021-9

Guidelines for condition assessment of educational facilities. Version 1 (2019). Department of Basic Education, Republic of South Africa

Gupta, G., Khan, M.A., Butola, R. & Singari, R. M. (2021). Advances in applications of Non-Destructive Testing (NDT): A review. *Advances in Materials and Processing Technologies.* https://doi.org/10.1080/2374068X.2021.1909332.

Hammer, J., & Pivo, G. (2016). The triple bottom line and sustainable economic development theory and practice. *Economic Development Quarterly.* https://doi.org/10.1177/0891242416674808.

https://www.europarl.europa.eu/news/en/headlines/economy/20151201STO05603/circular-economy-definition-importance-and-benefits.

Humberto, E. G., Wen-Chiao, L., & Semyon, M. M. (2012). A resilient condition assessment monitoring system. https://core.ac.uk/download/pdf/190695757.pdf

ICAO—The New Global Reporting Format for Runway Surface Conditions.

ISO 17359, Condition monitoring and diagnostics of machines—General guidelines.

ISO 13373, Mechanical vibration and shock—Vibration condition monitoring of machines.

ISO 13379, Data interpretation and diagnostic techniques which use information and data related to the condition of a machine.

ISO 13381, Condition monitoring and diagnostics of machines—Prognostics.

ISO 18436 Part II, Accreditation of organizations and training and certification of personnel—Part II—General requirements for training and certification—Vibration Analysis.

ISO 18436 Sub Parts under development for Oil Analysis and Thermal Imaging.

Kirchherr, J., Reike, D., & Hekkert, M. (2017). Conceptualizing the circular economy: An analysis of 114 definitions. *Resources, Conservation and Recycling, 127*, 221–232. ISSN 0921-3449. https://doi.org/10.1016/j.resconrec.2017.09.005.

Levy, D., Brown, H. S., & de Jong, M. (2010). The contested politics of corporate governance: The case of the global reporting initiative. Management and Marketing Faculty Publication Series. Paper 1. http://scholarworks.umb.edu/management_marketing_faculty_pubs/1

Marcia, L. B., Ivo, P. de M., Joao, P. M., Cairo, L. N., Helmut, P., & Elsa, H. (2017). Aircraft on-condition reliability assessment based on data-intensive analytics. In 2017 IEEE aerospace conference. IEEE Xplore https://doi.org/10.1109/AERO.2017.7943870.

Marlow, D., & Burn, S. (2008). Effective use of condition assessment within asset management. *Journal–American Water Works Association, 100*(1), 54–63. https://doi.org/10.1002/j.1551-8833.2008.tb08129.x.

Mateljak, Ž., & Mihanović, D. (2016). Operational planning level of development in production enterprises in the machine building industry and its impact on the effectiveness of production.

Economic   Research-Ekonomska   Istraživanja,   29(1),   325–342.   https://doi.org/10.1080/ 1331677X.2016.1168041.

Mayo, G., & Karanja, P. (2018). Building condition assessments—Methods and metrics. *Journal of Facility Management Education and Research, 2* (1), 1–11. https://doi.org/10.22361/jfmer/91666

Neumann, P., Kihlberg, S., Medbo, P., Mathiassen, S. E., & Winkel, J. (2002). A case study evaluating the ergonomic and productivity impacts of partial automation strategies in the electronics industry. *International Journal of Production Research, 40*(16), 4059–4075. https://doi.org/10.1080/00207540210148862.

Paik, J. K., & Melchers, R. E. (eds). (2008). In Condition assessment of aged structures. CRC Press.

Pellegrino, J. L., & Carole, T. M. (2004). Impacts of condition assessment on energy use: Selected applications in chemical processing and petroleum refining. Report prepared for U.S. Department of Energy Industrial Technologies Program Washington, D.C

Renwick, D. V., & Rotert, K. (2019). Deteriorating water distribution systems can impact public health. *Opflow, 45*(9), 12–15. https://doi.org/10.1002/opfl.1246

Sanchez-Silva, M., Klutke, G-A., & Rosowsky, D. V., (2011). Life-cycle performance of structures subject to multiple deterioration mechanisms. *Structural Safety, 33*(3), 206–217. ISSN 0167-4730. https://doi.org/10.1016/j.strusafe.2011.03.003.

Slaper, T. F., & Hall, T. J. (2011). The triple bottom line: What is it and how does it work? *Indiana Business Review, 86*(1)

Stankovic, M., Hasanbeigi, A., & Neftenov, N. (2020). In Marcello, B., Anamaría, N., Raphaëlle, O. (Eds.), Use of 4IR technologies in water and sanitation in Latin America and the Caribbean. IDB-TN-1910 Inter-American Development Bank.

The Maintenance Framework. Global Forum on Maintenance and Asset Management (GFMAM) 2016. ISBN: 978–0-9870602-5-9.

Verma, S. K., Bhadauria, S. S., Akhtar, S. (2013). Review of nondestructive testing methods for condition monitoring of concrete structures. *Journal of Construction Engineering, 2013*, Article ID 834572, 11p. https://doi.org/10.1155/2013/834572.

Vuchic, V. R. (2005). Urban transit: Operations, planning and economics. Wiley. ISBN 0–471-63265-1

Wahida, R. N., Milton, G., Hamadan, N., Lah, N. M. I. B. N., & Mohammed, M. A. H. B. (2012). Building condition assessment imperative and process. *Procedia—Social and Behavioral Sciences, 65*, 775–780 https://doi.org/10.1016/j.sbspro.2012.11.198.

Asset Retirement Stage

8

Abstract

The discourse in this chapter is focused on issues concerning the retirement stage in the life of an engineered asset. The emphasis here is how to implement the asset 'end-of-life' recommendations arising as a consequence of condition and performance assessments. Subsequent to condition monitoring and assessment, the initial part of the decision-making is to ascertain whether or not an asset has reached end-of-life, and if so, what the next course of action should be. Additional issues to be considered with the end-of-life recommendations include statutory requirements (e.g., environmental impact), contractual obligations (especially those that arise from the acquisition options), and valuation of the asset.

8.1 Asset 'End-of-Life' Definitions

A conundrum that is fairly common arises from the various definitions of asset life. Some of the end-of-life definitions[1] include:

---

[1]Animah, I., & Shafiee, M. (2017). Condition assessment, remaining useful life prediction and life extension decision making for offshore oil and gas assets. *Journal of Loss Prevention in the Process Industries*. https://doi.org/10.1016/j.jlp.2017.04.030.

Okoh, C., Roy, R., Mehnen, J., & Redding, L. (2014). Overview of remaining useful life prediction techniques in through-life engineering services. *Procedia CIRP, 16*, 158–163.

Ruitenburg, R. J, Braaksma, A. J. J., & Van Dongen, L. A. M. (2014). A multidisciplinary, expert-based approach for the identification of lifetime impacts in asset life cycle management. *Procedia CIRP, 22*, 204–212.

Sahu, A. K, Narang, H. K, Sahu, A. K., & Sahu, N. K. (2016). Machine economic life estimation

© The Author(s), under exclusive license to Springer Nature Switzerland AG 2021   113
J. E. Amadi-Echendu, *Managing Engineered Assets*,
https://doi.org/10.1007/978-3-030-76051-9_8

*Design life is the expected total time that an asset should provide the capability of being utilised towards the realization of the intended value for the user. Design life is a theoretical life of an asset that is conventionally estimated by designers, manufacturers, and suppliers. The estimation is based on material selection and product design standards, as well as anticipated operating conditions, *et cetera*.

*Service life is the period of time during which a deployed asset can deliver the levels of service demanded by the user. Service life strongly depends on the demands placed on an asset, all other factors considered.

*Useful life is the period of time that an asset remains in use as deployed.

*Economic life is the period of time during which an asset provides economic profit or financial benefit.

*Remaining life—Often, the primary reason for conducting condition and performance assessments is to determine the remaining or residual life of an asset. If the assessments indicate that an asset is underperforming but still functioning as deployed, then remaining life is considered as the period of time after the assessments until functional failure of the asset.

*According to the Oxford dictionary, *residual* means"… *remaining* after other things have been subtracted or allowed for…"In this regard, residual life is often considered as the difference between the time when the asset is retired and the estimated or theoretical life of an asset.

8.2   Asset 'End-of-Life' Considerations

Funding and financing, technology, environment/ecology, and socio-economic and socio-political considerations also feature during the retirement stage in the life of an asset. Evaluations of the benefit-to-cost ratio, safety concerns, increasing risk of functional failure, diminishing economic profit, technological obsolescence and diminishing financial valuation are significant considerations that influence decisions to retire an asset.

based on depreciation-replacement model. *Cogent Engineering, 3.* http://dx.doi.org/10.1080/23311916.2016.1249225.

Searles, C., & Schiemann, M. (2014). Understanding and differentiating design life, service life, warranty and accelerated life testing for lead acid batteries, pp. 1–9.

Thombre, S. C. & Kotwal, R. (2015, March). Assessment of residual life of engineering component. *International Journal of Engineering Research and Technology JETIR, 2*(3) (ISSN-2349–5162).

Mikhail, V., Anton, K., Anna, V., & Vera, M. (2016). Existing models residual life assessment of structures and their comparative analysis. *Procedia Engineering, 165,* 1801–1805.

Notwithstanding the evidential logic and epistemology of the decision-making process, economic value coupled with safety and environmental risks tend to be highly prioritised irrespective of how the life of an asset is defined. Statutory requirements concerning asset end-of-life decisions usually focus on safety and environmental risks associated with, for example, ageing phenomena.[2] The regulations often stipulate how asset deterioration and performance degradation should be monitored and assessed.

In many instances, the end-of-life decisions are strongly influenced by the option chosen to acquire an asset. There are obligations implied in any option chosen to acquire an asset. Such obligations are typically stated in clauses embedded within the terms and conditions of the acquisition contract. It is remarkable that guarantee and/or warranty clauses tend to be muted but take on renewed significance when a decision has been taken to retire an asset. It is not uncommon that many practitioners suddenly find out about the obligations when they have to implement the decision to retire an asset. This is particularly so for long-lived assets, and nowadays, the rapidity of technology obsolescence tends to exacerbate the challenge even for a short-lived asset.

The prevalence of rapid evolutions in technology means that technological obsolescence[3] often overrides other considerations. The effect is particularly pronounced where the current component, spare part, subassembly, subsystem or even the asset is not only superseded by a new model but also, the legacy version is no longer available. "Obsolescence affects all products and it impacts upon all stages of their life", hence, the management of obsolescence is increasingly becoming a specialised topic in asset management.[4]

---

[2]Barrie, H. (1994, October). Europe's ageing infrastructure: Politics, finance and the environment. *Utilities Policy, 4*(4), 243–252.

Klaas van, B. (2017). Societal burden and engineering challenges of ageing infrastructure. *Procedia Engineering, 171*, 53–63.

André, O. (2013). Final report on optimization of management strategies under different traffic, climate change and financial scenarios. CEDR Call (2013). Ageing Infrastructure Management—Understanding Risk Factors.

[3]Amankwah-Amoah, J. (2017). Integrated versus add-on: A multidimensional conceptualisation of technology obsolescence. MPRA Paper No. 86353, posted 24 Apr 2018 08:19 UTC. https://mpra.ub.uni-muenchen.de/86353/.

Romero Rojo, F. J., Roy, R., & Kelly, S. (2012). Obsolescence risk assessment process best practice. *Journal of Physics: Conference Series, 364*, 012–095.

Sandborn, P. (2007, December). Software obsolescence—Complicating the part and technology obsolescence management problem. *IEEE Trans on Components and Packaging Technologies, 30*(4), 886–888.

[4]IEC 62,402:2007: Obsolescence management—application guide.

Expert Group 21: Obsolescence Management; EDSTAR—European Defence Standards Reference System.

8.3   Asset Retirement Options

The decision that an asset has reached its end-of-life can lead to a number of courses of actions. The first and foremost action is to terminate the use of the asset. As trivial as this action may seem, it can pose some interesting challenges given that there are various stakeholders to an asset. In fact, the action to terminate the use of an asset can become embroiled in legal contestation. Some of the retirement options pose at least two sets of interesting dilemmas. The first is that the recommended retirement activities have to be implemented in either green-field, brown-field or mixed brown/green environments. The second is that, if a component, spare part, subassembly, or subsystem is to be retired, then, the need might arise to justify whether or not the associated expenditure can be capitalised.

The flow diagram in Fig. 8.1 provides a pictorial summary of the options that can be implemented during the retirement stage in the life of an asset.

8.3.1   Mothballing an Asset

In some instances, the recommendation may be to 'mothball' an asset, that is, to place the asset in preservation so that it can be utilized in the future.[5] Mothballing or temporary retirement includes terminating the use of an asset even though the reason for deploying the asset still prevails. Therefore, mothballing is not strictly because an asset has truly reached end-of-life but, often, the predominant reason for discontinuing the use of the asset could be that the asset becomes stranded as a consequence of several factors, e.g., lack of funds/finance, withdrawal of operating licence, regulatory prohibition, challenges due to climate change, *et cetera*.

In deciding to place an asset in temporary retirement, it is vital to ensure that the asset can be readily returned to service at a later date, thus, the paramount requirement is that an asset must be preserved in a serviceable condition whilst the asset is in a mothballed state. Mothball preservation of an asset involves, *inter alia*, condition monitoring to identify, track and remedy ageing deterioration so as to maintain asset integrity and to guarantee operational readiness.

[5]Scott, P. (2004). Asset preservation: Mothballing and lay-up. https://doi.org/10.1115/POWER2004-52053.

Ben, C., & Jeremy, M.D. (2014). Stranded generation assets: Implications for European capacity mechanisms, energy markets and climate policy. Stranded Assets Programme Working Paper. Smith School of Enterprise and the Environment, University of Oxford.

Suphi, S., Marie-Theres von, S. (2019). Climate policy, stranded assets, and investors' expectations. *Journal of Environmental Economics and Management, 100*, March 2020, 102–277. https://doi.org/10.1016/j.jeem.2019.102277.

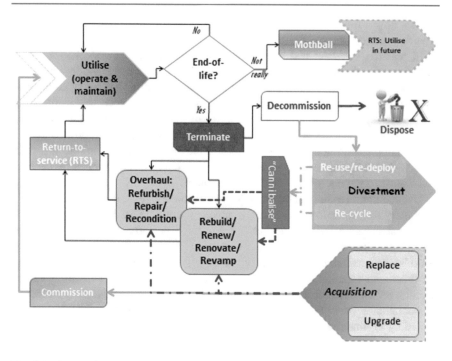

Fig. 8.1 Asset retirement options

8.3.2 Replace, Upgrade, Overhaul and Repair

Strictly speaking, replacement means that the life of the item being replaced has ended. Replacement has become the most common end-of-life option especially for components, spare parts and sometimes subassemblies. Technology has long advanced to the point where most items at the component and spare part levels of the asset hierarchy are no longer repairable. For such items, the focus is now shifting to recycling the composite materials.

Upgrade typically involves replacing part(s) of the legacy asset with new parts that feature superior functionality and performance; it may also mean replacing an entire asset with another asset that features superior functionality and performance. Both replacement and upgrade imply the acquisition of something new. In instances where the whole asset is replaced or upgraded, the new asset has to be commissioned. The commissioning tends to be relatively less challenging if the new asset is ring-fenced as a green-field project. Otherwise, the commissioning is constrained by the brown-field nature of the replacement or upgrade project.

Traditionally, repair was the most common maintenance activity performed on an asset. Perhaps, one reason is that repairing an item is perceived to be relatively less expensive than replacing the item. Another muted reason is that repairing provides an opportunity to learn how an item is put together, after all, 'fixing teaches how a thing works'. Furthermore, there is a controversial perception that repairing reduces waste.

In practice, overhaul and repair are usually considered as asset life-extension activities even though the actions may involve replacing and upgrading elements that result in an asset with increased functional capability. By convention, overhauling involves stripping or taking apart an asset in order to examine and implement repairs where necessary. In terms of the option to overhaul and repair, the presumption is that the asset is repairable, even if, depending on how an asset is defined, implementing the repair may involve the replacement of components, spare parts, subassemblies, or subsystems that may have failed or may have been ascertained to have a higher likelihood of failure.

Overhaul and repair usually describe a situation where an item of equipment or machinery can be restored to a 'serviceable condition'. That is, restoring existing functional capacity so that the item continues to perform within the prescribed operating regime(s) and for the purpose that is was originally deployed. Similarly, refurbish and recondition are regarded as activities that involve the restoration of the functional capability so that an asset can be redeployed to perform the same function for the same or a different purpose. The cost of restoration may be treated as operational or capital expenditure depending on how the asset is defined according to financial recognition and accounting rules.

8.3.3   Rebuild, Renew, Renovate, Revamp, Etc

Renew, renovate, and revamp are activities that typically encompass replace, upgrade, overhaul and repair, and they generally refer to situations concerning the restoration and return-to-service of assets commonly grouped as facilities and infrastructure. Curiosity arises as to whether the original asset still exists given that renewing, renovating, or revamping also include overhaul, repair, replace, and upgrade activities. The paradox arises because equipment and machinery usually form part and parcel of facilities and infrastructure assets (re: cyber physical systems). Whether or not the cost of rebuilding, renewing, renovating, or revamping is capitalised depends on how the asset is defined and configured in conformity with the rules for financial recognition, accounting, and reporting on assets. In instances where these actions are more than restorative and lead to improved/increased functional capability, it is tantalising to argue that the corresponding expenditure should be capitalised with belief that the asset's life has been extended!

8.3.4   Disposal and Divestment

If the decision is to permanently terminate and decommission the use of the asset, then the options include disposal and/or divestment of the asset in a manner that is safe, ethically acceptable and environmentally permissible. Decommissioning effectively means that the asset is no longer useful for the purpose(s) of the original investment. The implication is really that the life of the asset has truly ended. It is also in this regard that asset replacement has its true meaning—the decommissioned asset is replaced by acquiring and commissioning a new asset to perform the same function, notwithstanding the fact that the new asset may be functionally superior to the one that was replaced. A dilemma often arises with regard to upgrades, especially where the resulting asset can also be utilised outside of the stated legacy operating regime(s), or where the upgrade makes it functionally possible to deploy the asset for other purposes pre-specified as per the investment decision. Remarkably, the rules for financial recognition accounting and reporting generally favour the capitalization of true asset replacements and technologically superior upgrades.

Environmental concerns about pollution and waste mean that outright disposal of an asset has become prohibitive. Instead, the preference is to find alternative ways of retaining an asset within the confines of the socio-economic domain—that is the primary goal of the circular economy model. In practice and where feasible, an asset may be decommissioned so that it can be 'cannibalised', that is, by salvaging the functioning components, spare parts, subassemblies or even subsystems of the asset so that they can be re-used to perform the same functions within a similar asset elsewhere. In many instances, cannibalisation[6] not only facilitates incremental and concurrent adaptation to new technologies but also, it can be positively considered as a way of adopting and assimilating disruptive technological innovations.

Divestment provides a pragmatic option towards compliance with environmental regulations, especially given that globalised markets and supply chains have become common in the era of *Society 5.0*, facilitating tighter linkages between manufacturers, vendors, suppliers, systems integrators, *et cetera*, with asset users. Furthermore, globalised standardisation of components, spare parts, subassemblies, and subsystems coupled with advances in modularised designs means that, on the one hand, an asset can be completely-built-up (CBU), and, on the other hand, an asset can be completely-knocked-down (CKD).[7] Thus, in situations where an asset cannot be sold-off as a whole, then the asset can be dismantled to recover components, spare parts, subassemblies, and subsystems that have saleable or tradable value. For very long lived assets that have been in existence well before CBU/CKD and modular design prevailed, or assets that cannot be decomposed into re-useable components, spare parts, subassemblies, and subsystems, the last option is to recycle,[8] that is, to destroy the asset in order to extract materials that can be used to create new assets.

---

[6]Rui, Z., & Ahmed, G. (2015). Detailed cannibalization decision making for maintenance systems in the military context. https://cradpdf.drdc-rddc.gc.ca/PDFS/unc214/p803151_A1b.pdf.

Nuno de Oliveira, A. A, & Ghobbar. (2010). Cannibalization: How to measure and its effect in the inventory cost. In 27th International Congress of the Aeronautical Sciences. http://www.icas.org/ICAS_ARCHIVE/ICAS2010/PAPERS/319.PDF.

[7]Anastasia, M. (2019). Cost efficiency and waste reduction in completely knocked down production. Thesis Report. Department of Industrial Engineering. Technical University of Jonkoping.

[8]Lingen, Z., Zhenming, X. (2016). A review of current progress of recycling technologies for metals from waste electrical and electronic equipment. *Journal of Cleaner Production, 127*, 19–36.

8.4 Summary of the Chapter

The discourse in this chapter included brief definitions of asset life and issues to be considered when implementing recommendations of end-of-life options. The activities that are commonly carried out during the asset retirement stage have been discussed in reasonable detail.

8.5 Exercises

1. Explain why it is important to define asset life. Choose an asset and discuss how the definitions of asset life apply to your chosen asset. On what basis will you decide that the asset has reached its end-of-life?
2. Explain in detail how you would retire your chosen asset following a decision that the asset has reached its end-of-life.
3. Is mothballing an asset justifiable nowadays? Explain your reasoning based on a case study.

References and Additional Reading

Amankwah-Amoah, J. (2017). Integrated versus add-on: A multidimensional conceptualisation of technology obsolescence. MPRA Paper No. 86353, posted 24 Apr 2018 08:19 UTC. https://mpra.ub.uni-muenchen.de/86353/.

Anastasia M. (2019). Cost efficiency and waste reduction in completely knocked down production. Thesis Report. Department of Industrial Engineering. Technical University of Jonkoping.

André, O. (2016). Final report on optimization of management strategies under different traffic climate change and financial scenarios. CEDR Call 2013: Ageing Infrastructure Management-Understanding Risk Factors.

Animah, I., & Shafiee, M. (2017). Condition assessment remaining useful life prediction and life extension decision making for offshore oil and gas assets. *Journal of Loss Prevention in the Process Industries*. https://doi.org/10.1016/j.jlp.2017.04.030

Barrie, H. (1994). Europe's ageing infrastructure: Politics finance and the environment. *Utilities Policy, 4*(4), 243–252.

Ben, C. & Jeremy, M. D. (2014). Stranded generation assets: Implications for European capacity mechanisms, energy markets and climate policy. Stranded Assets Programme Working Paper. Smith School of Enterprise and the Environment, University of Oxford.

De Oliveira, N., & Ghobbar, A. A. (2010). Cannibalization: How to measure and its effect in the inventory cost. In 27th International Congress of the Aeronautical Sciences. http://www.icas.org/ICAS_ARCHIVE/ICAS2010/PAPERS/319.PDF.

Expert Group 21: Obsolescence Management, EDSTAR—European Defence Standards Reference System.

IEC 62402:2007 : Obsolescence management—Application guide.

Okoh, C., Roy, R., Mehnen, J., & Redding, L. (2014). Overview of remaining useful life prediction techniques in through-life engineering services. *Procedia CIRP, 16*, 158–163.

Paul, S. (2004). Asset preservation: Mothballing and lay-up. https://doi.org/10.1115/POWER2004-52053

Romero Rojo, F. J., Roy, R., & Kelly, S. (2012). Obsolescence risk assessment process best practice. *Journal of Physics: Conference Series, 364*, 012095.

Rui, Z., & Ahmed, G. (2015). Detailed cannibalization decision making for maintenance systems in the military context. https://cradpdf.drdc-rddc.gc.ca/PDFS/unc214/p803151_A1b.pdf.

Ruitenburg, R. J., Braaksma, A. J. J., & Van Dongen, L. A. M. (2014). A multidisciplinary expert-based approach for the identification of lifetime impacts in asset life cycle management. *Procedia CIRP, 22*, 204–212.

Sahu, A. K, Narang, H. K , Sahu, A. K. & Sahu, N. K. (2016). Machine economic life estimation based on depreciation-replacement model. *Cogent Engineering, 3*. https://doi.org/10.1080/23311916.2016.1249225.

Sandborn, P. (2007, December). Software obsolescence—Complicating the part and technology obsolescence management problem. *IEEE Trans on Components and Packaging Technologies, 30*(4), 886–888.

Searles ,C., & Schiemann, M. (2014). Understanding and differentiating design life service life warranty and accelerated life testing for lead acid batteries, pp. 1–9.

Suphi, S., Marie-Theres von, S. (2019). Climate policy stranded assets and investors' expectations. *Journal of Environmental Economics and Management, 100*, 102277. https://doi.org/10.1016/j.jeem.2019.102277.

Thombre, S.C., & Kotwal, R., (2015). Assessment of residual life of engineering component. *International Journal of Engineering Research and Technology JETIR March 2015, 2*(3). (ISSN-2349–5162).

van Breugel, K. (2017). Societal burden and engineering challenges of ageing infrastructure. *Procedia Engineering, 171*(2017), 53–63.

Volkov, M., Kibkalo, A., Vodolagina, A., & Murgul, V. (2016). Existing models residual life assessment of structures and their comparative analysis. *Procedia Engineering, 165*(2016), 1801–1805.

Zhang, L., & Zhenming, X. (2016). A review of current progress of recycling technologies for metals from waste electrical and electronic equipment. *Journal of Cleaner Production, 127*(2016), 19–36.

SAMP & EAMBoK

9

Abstract

The discourse in this chapter briefly highlights the pertinent issues of strategic asset management planning and engineering asset management body of knowledge (EAMBoK).

9.1 Engineering Asset Management Strategy and Policy

For the purposes of this book, strategy can be interpreted as a careful articulation of how to direct actions in order to achieve desired objectives within a specified time frame. The time frame of a strategy could be short, medium or long term. Thus, there are two parts to a strategy. The first part is articulating or setting the direction, i.e., defining the objectives to be achieved over a specified time frame; and the second part is prescribing actions intended towards achieving the desired objectives within the specified time frame. Many pundits regard the first part as strategy formulation, and the second part as strategic planning.

Policy can be interpreted as a set of rules that guide the implementation of strategy. It is paramount that the set of rules are unambiguous and consistent, and that the rules are adopted to guide how the prescribed actions will be directed towards achieving the desired objectives within the specified time frame. The combination of the prescribed actions and rules may be formalised as official procedures for strategy implementation. In some organisations, certain policies get transformed into intransigent cultural norms that sometimes make it difficult to articulate and implement new strategies. Interestingly, in situations where there is no distinction between strategy and policy, it should not be surprising that policy and strategy may become misaligned.

9.1.1   Asset Management Strategy

The prime objective is to realise value, hence an asset management strategy provides direction for management actions to result in the realisation of value from an asset over a specified time frame. The management actions can be concatenated in terms of the respective life stages of an asset. Thus, it is conceivable that an overall asset management strategy can be concatenated into (i) asset acquisition strategy, (ii) asset utilisation strategy, and (iii) asset retirement strategy. Similarly, the sequencing and linking of the management actions result in the respective processes for acquiring, utilising, and retiring an asset.

For practical purposes, a generic asset management strategy should include:

- unambiguous and consistent set of objectives to be achieved within a specified time frame;
- the scope of management actions in terms of prescribed work processes, resources and proposed schedules; and
- formalised rules to ensure that the work processes and resources are directed towards achieving the objectives with the specified time frame.

9.1.2   Strategic Asset Management Plan

There are many definitions of a strategic plan despite the fact that strategic planning[1] is well-established in management practice. This book adopts the definition of strategic planning as a 'deliberative, disciplined effort to produce fundamental decisions and actions that shape and guide what an organization (or other entity) is, what it does, and why'.[2] In management parlance, SWOT analysis is conventionally applied and the strategic planning process is typically conducted in a group setting. The strategic planning process requires a facilitator, preferably a moderator who is not a formal member of the group.

The reality is that supervening circumstances causes strategic objectives to change even within the specified time frame. Hence, the general aim of a strategic planning exercise is to ensure that management actions and rules are aligned with circumstantial changes to strategic objectives. Since every socio-economic/political

---

[1]Ida, S., Azahari, B. R., Munauwar, B. M., & Rushami, Z. B. Y. (2015). Strategic planning and firm performance: a proposed framework. *International Academic Research Journal of Business and Technology, 1*(2) 201–207.

Arasa, R., & K'Obonyo, P. (2012). The relationship between strategic planning and firm performance. *International Journal of Humanities and Social Science, 2*(22) [Special Issue–November 2012]

Mintzberg, H. (1993). The pitfalls of strategic planning. *California Management Review Fall , 36* (1), 32 ABI/INFORM Global

Steiner, G. A. (1979). Strategic planning. The Free Press. ISBN 0–02–931110–1.

[2]Bryson, J. M., Edwards, L. H., & Van Slyke, D. M. (2018) (eds). Toward a more strategic view of strategic planning research. *Public Management Review, 20*(3).

entity or organisation depends on assets to conduct business, the implication is that strategic planning encompasses asset management. Strategic planning in the private sector is about how to manage engineered assets plus human and other resources to achieve the overarching strategic objective of profit. Strategic asset management planning (SAMP) is a slightly narrower scope than strategic planning in the sense that SAMP focuses on the management of assets within the constraints of available human capital and other complementary resources.

Nowadays, SAMP generally refers to strategic planning in the context of public sector organisations.[3] In most countries, the public sector is the owner, custodian and manager of the largest base of assets (re: infrastructural facilities for education, public health, water and sanitation, state security, transportation, *et cetera*). Hence, SAMP is vital for the public sector, especially as managers have to concurrently deal and contend with fiscal prudence in budgetary execution, economic efficiency in terms of resource allocation, environmental compliance, and social equity in service delivery. The management of assets[4] in the public sector depends on country-specific governance structures, policy, legislative and regulatory regimes, as well as the level of economic, political and social development. Furthermore, external macro-economic and socio-political environments oftentimes exert conflicting and contradictory influences on the management of public sector assets. A major implication of these challenges is that public–private partnership, outsourcing and servitisation arrangements should form part and parcel of SAMP.

9.2 Engineering Asset Management Body of Knowledge (EAMBoK) Issues

It is arguable that the formalisation of engineering asset management as a multidisciplinary area of knowledge and practice has accelerated following the release of the first set of ISO 5500× series of standards. The consequence is that the establishment of a comprehensive body of knowledge (BoK) for education, training, research, skills development, and practice is lagging. As at the publication of this book, there is not a global and generally accepted body of knowledge for the management of engineered assets, despite the fact that there are several relevant publications by a growing number of professional associations and fraternities.

The excitement following the release of the first set of ISO 5500× standards has spurred a trend to recast legacy programmes in reliability and maintenance as part of educational curricula leading to asset management qualifications. Maintenance

---

[3]Elbanna, S., Andrews, R., & Pollanen, R. (2016). Strategic planning and implementation success in public service organisations: Evidence from Canada. *Public Management Review, 18*(7), 1017–1042. https://doi.org/10.1080/14719037.2015.1051576

[4]Amadi-Echendu J. (2015) Assessment of engineering asset management in the public sector. In: P. Tse, J. Mathe, K. Wong, R. Lam, C. Ko C. (Eds.), Engineering asset management—systems, professional practices and certification. Lecture Notes in Mechanical Engineering. Springer, Cham. https://doi.org/10.1007/978-3-319-09507-3_97

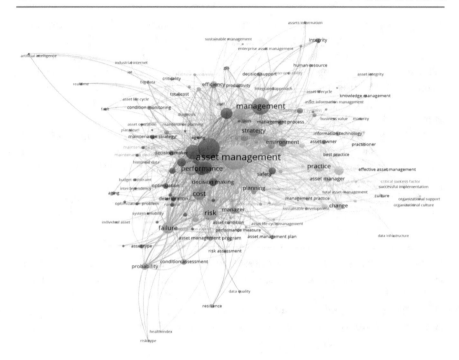

Fig. 9.1 A visualised mapping of research topics in asset management {Ref: Gavrikova [(Gavrikova et al. 2020)]}

functions in organisations are also being recast into asset management departments, and hitherto maintenance associations are transforming into asset management societies. Despite the fact that the series is still being developed, many business organisations are being audited and certified in accordance to the first set of ISO 5500× management system standards. Some of the established maintenance associations that have transformed into asset management fraternities are issuing professional certificates to invdividuals based on proprietary frameworks that are purportedly ISO 5500× compliant, even though more standards are still being developed under the banner of ISO 5500× series. It is intriguing that the 'ISO 5500× excitement' has not yet resulted in a global and generally accepted BoK. There is also the need to found an accreditation framework for asset management qualifications and professional certification.

The mapping in Fig. 9.1[5] indicates there is scope for recasting legacy courses into new curricula and academic programmes that embrace the multidisciplinary nature of asset management. In as much as academic programmes may be heterogeneous to reflect regional and local preferences, however, it is important that asset management curricula should be based on a generally accepted body of

---

[5]Gavrikova, E., Volkova, I., & Burda, Y. (2020). Strategic aspects of asset management: An overview of current research. *Sustainability*, *12*(15), 5955. https://doi.org/10.3390/su12155955

knowledge. It is envisaged that the growing number of associations, institutes, and societies will facilitate the evolution towards normative frameworks for (i) education and qualifications accreditation; (ii) skills and competency development; and (iii) professional certification in asset management.

9.3   Summary of the Chapter

The highly pertinent topic of strategic asset management planning has been introduced. The discourse also highlighted the importance of a generally accepted body of knowledge as a basis for education and qualifications accreditation, skills and competency development, as well as professional certification in asset management.

9.4   Exercises

1. Assume that you have been appointed as an asset manager of an organisation in any public or private sector of industry. Choose an imaginary organisation and an industry sector (e.g., agriculture and agro-allied, education and research, health, manufacturing, mining and minerals processing, security, transportation, etc.). Explain, in convincing detail, how you would go about developing a strategic asset management plan for the organisation.
2. With regard to managing engineered assets, discuss some of the challenges confronting the development of curricula for:

   a. Education and qualifications accreditation;
   b. Skills and competency development; and
   c. Professional certification.

References and Additional Reading

Amadi-Echendu J. (2015). Assessment of engineering asset management in the public sector. In P. Tse, J. Mathew, K.Wong, R. Lam, C. Ko (Eds.), Engineering asset management—systems, professional practices and certification. Lecture Notes in Mechanical Engineering. Springer, Cham. https://doi.org/10.1007/978-3-319-09507-3_97

Arasa, R., & K'Obonyo, P. (2012). The relationship between strategic planning and firm performance. *International Journal of Humanities and Social Science, 2*(22) [Special Issue– November 2012]

Bryson, J. M., Edwards, L. H., & Van Slyke, D. M. (eds). (2018). Toward a more strategic view of strategic planning research. *Public Management Review, 20*(3)

Elbanna, S., Andrews, R., & Pollanen, R. (2016). Strategic planning and implementation success in public service organisations: Evidence from Canada. *Public Management Review, 18*(7), 1017–1042. https://doi.org/10.1080/14719037.2015.1051576

Gavrikova, E., Volkova, I., & Burda, Y. (2020). Strategic aspects of asset management: An overview of current research. *Sustainability, 12*(15), 5955. https://doi.org/10.3390/su12155955

Ida, S., Azahari, B. R., Munauwar, B. M., Rushami, Z. B. Y. (2015). Strategic planning and firm performance: A proposed framework. *International Academic Research Journal of Business and Technology, 1*(2) 201–207

Mintzberg, H. (1993). The pitfalls of strategic planning. *California Management Review Fall ,36* (1), 32 ABI/INFORM Global

Steiner, G. A. (1979). In *Strategic planning*. The Free Press. ISBN 0–02–931110–1

Printed in the United States
by Baker & Taylor Publisher Services